BUSINESS REPLY CARD

FIRST CLASS PERMIT NO. 442 FLUSHING, N.Y.

POSTAGE WILL BE PAID BY ADDRESSEE

**American Map Corp.
46-35 54th Road,
Maspeth, N Y 11378**

BUSINESS REPLY CARD

FIRST CLASS PERMIT NO. 442 FLUSHING, N.Y.

POSTAGE WILL BE PAID BY ADDRESSEE

**American Map Corp.
46-35 54th Road,
Maspeth, N Y 11378**

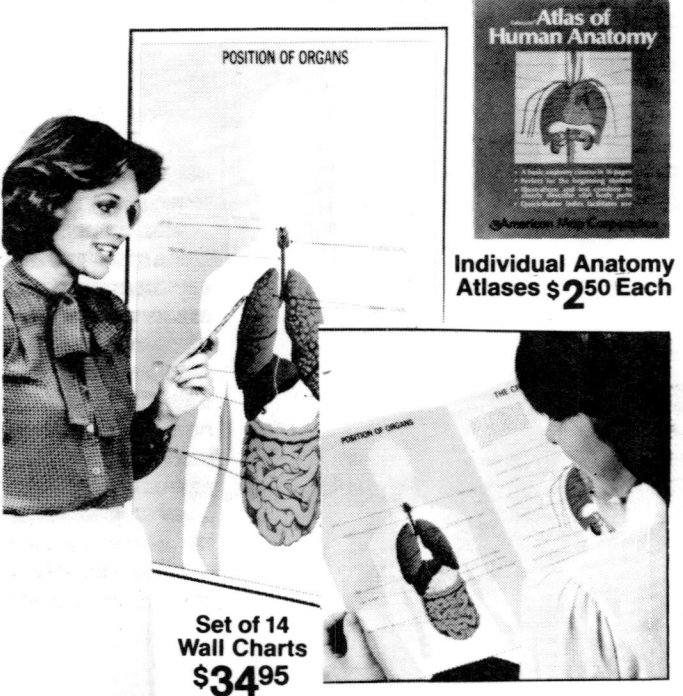

Anatomical Feature	Chart number, bold / Feature number, light
U	
ULCER	
Duodenal	**6**-A6
Peptic	**6**-D2
Ulna	**1**-26;**19**-96
Umbilical Cord	**11**-H7,K5;**12**-C6
Ureters	**10**-A10,D18;**12**-A1;**13**-A3,B1
	19-98;**21**-109
URETHRA	
External Meatus	**13**-A15;
Membranous	**13**-A10,B8
Penile	**13**-A13
Prostatic	**13**-A8
Vaginal	**12**-A10
URINARY BLADDER	
Urinary Bladder	**10**-D13;**12**-A8;**13**-A4;**19**-86;**21**-103;**31**-E12
Apex	**13**-A6c
Body	**13**-A6a
Neck	**13**-A6b
UTERUS	
Uterus	**11**-J1
Corpus	**12**-A21
Endometrium	**11**-J16
Orifice of	**12**-A18
Os of, External	**11**-J12;**12**-C3
Inner	**12**-C2
Outer	**11**-J10;**12**-C3;**11**-J12
Rectouterine Pouch (of Douglas)	**12**-A20
Section, after Delivery	**12**-C
Uterus, Fallopian Tubes, and Ovaries	**11**-J
Uvula	**20**-III2
V	
Vagina	**11**-J13;**12**-C4,A11,A12
Vagus Nerve	**8**-A1,D5;**26**-21,26;**27**-A13,
	30-C1
VALVES OF THE HEART	
Aortic	**4**-A14,C1
Posterior Cusp	**4**-C7a
Right Cusp	**4**-C7b
Left Cusp	**4**-C7c
Mitral	**4**-A16,C3
Posterior	**4**-C3a
Anterior	**4**-C3b
Pulmonary	**4**-A13,C5
Anterior Cusp	**4**-C5a
Left Cusp	**4**-C5c
Right Cusp	**4**-C5b
Semilunar Cusps	
Closed	**4**-C1
Open	**4**-C2
Tricuspid (Atrio-Ventricular)	**4**-A13,A14,A27
Anterior Cusp	**4**-C8a
Medial Cusp	**4**-C8b
Posterior Cusp	**4**-C8c
Trabeculae Carnae	**4**-A25
Valves of Lymph Vessels	**24**-D
VEINS	
Arcuate	**10**-C10
Axillary	**19**-26
Basilic	**19**-35
Brachiocephalic	**4**-A3
Left	**19**-21
Right	**19**-113
Cephalic	**19**-32
Accessory	**19**-38
of Eye	**2**-7
Femoral	**19**-62,65
Lateral Circumflex	**19**-64
Gonadal	**10**-D20
Hepatic	**10**-D27
Iliac	**13**-A1;**19**-91
Interlobular, of Kidney	**10**-C7
Intestinal	**6**-C7
Jugular, External	**19**-119
Internal	**19**-20
Median Antebrachial	**19**-39
Median Cubital	**19**-36
Pulmonary	**4**-B3
Left	**4**-A11
Right	**4**-A31
Renal	**10**-A12,D22
VEINS (Continued)	
Ramification of	**10**-A6
Saphenous, Great	**19**,73,74
Of Skin	**3**-19,27,28
Stellate	**10**-C4
Subclavian	**19**-22
Thyroid, Inferior	**19**-116
Vena Cava	
Inferior	**4**-A24,B6;**10**-D25;**19**-102
Superior	**4**-A4,B2;**19**-111;**24**-29;**26**-35
Venule, Skin	**3**-17
Vas Deferens	**13**-A9,B16,B2,B4,B5
VENTRICLE	
Fourth (Brain)	**15**-10;**27**-D35
Lateral (Brain)	
Anterior Horn of	**27**-B2
Inferior Horn of	**27**-B19
Posterior Horn of	**27**-B15
Left (Heart)	**4**-A15,B8
Right (Heart)	**4**-A28,D8;**7**-20
Third (Brain)	**15**-17;**27**-B21,D10
Tela Choroidea of	**27**-D12
Venule	**3**-17
Vermis	**27**-B14,D38
Vertebral Column	**26**-23
VERTEBRAE	
Cervical	**1**-9;**21**-15
Lumbar	**21**-18
Fifth Lumbar	**1**-24
Thoracic	**21**-16
Twelfth Thoracic	**1**-19
VESICLE	
Vesiculouterine Pouch	**12**-A7
Germinal	**11**-G2
Seminal	**13**-A25,B6
VESSELS: See Blood, Vessels; Lymph Vessels	
Vestibule	
of Ear	**9**-B8
Fenestra of	**9**-A11
Villus (Villi) in Development of Human Embryo	**11**-F2,H3
Intestinal Villus	**6**-C;**25**-I
Artery and Vein	**25**-I2
Digestive Disorders	**6**
VISION	**2**;**26**-12
Calcarine Sulcus	**27**-D1
Functional Area of Brain	**26**-12
Hyperopia (Farsightedness)	**2**-C
With Glasses	**2**-C41
Without Glasses	**2**-C40
Line of Comparison of Differences in Focal Plane	**2**-1 A-C
Myopia (Nearsightedness)	**2**-B
With Glasses	**2**-B38
Without Glasses	**2**-B37
Normal Sight	**2**-A
Protective Mechanism of the Eye	**14**
Sight as a Function of the Brain	**15**
Visual Speech Area of Brain	**26**-9
VOCAL CORD	**7**-26;**8**-A6;**20**-IA5,IB5,IIB10;**25**-H6
Vestibular Fold	**20**-IA12,IB12
Laryngeal Polyp of	**20**-IC1
Vocal Slit	**20**
Viscera	**21**-101,109
W	
WALL	
Abdominal	**10**-D8
Orbital	**2**-A3
White Matter	**27**-C7
X	
Xiphoid Process	**1**-18;**19**-103
Y	
Yolk of Human Egg	**11**-G1,H8
Z	
Zygomatic Arch	**1**-6;**14**-10;**26**-43
Zygomatic Bone	**19**-13
Zygote	**11**-A,B,C,D
Zona Pellucida	**11**-G4

Anatomical Feature	Chart number, bold Feature number, light	Anatomical Feature	Chart number, bold Feature number, light
MOUTH (Continued)		NERVES (Continued)	
Mouth, Functional	**26**-52	Cervical	**21**-83
Frontal Section through	**20**-III	Cervical Plexus	**21**-87
MUSCLES		Cochlear	**9**-22
Muscles	**21**	Cross Section of	**3**-18
Adductor Magnus	**21**-72	Facial (VII)	**27**-A24
Arrector Pili	**3**-13	Glossopharyngeal (IX)	**27**-A20
Biceps Femoris	**21**-70	Hypoglossal (XII)	**27**-A14
Brachialis	**21**-57	Infraorbital Foramen	**14**-15
Brachioradialis	**21**-58	Intercostal	**8**-A10
Ciliary	**2**-19	Intermediate Nerve	**27**-A22
Deltoid	**21**-55	Lateral Femoral Cutaneous	**21**-96
Dorsal Interossei (hand)	**21**-62	Lumbar	**21**-85
Extensor Carpi Radialis	**21**-60	Medial Brachial Cutaneous	**21**-95
Extensor Carpi Ulnaris	**21**-61	Median	**21**-90
Extensor Digitorum	**21**-59	Motor	**27**-C12
Flexor Digitorum Longus	**21**-81	Nervous Control of Respiration	**8**
Gastrocnemius	**21**-78	Oculomotor (III)	**27**-A28,D34
Geniohyoid	**5**-7; **20**-IIB19	Optic (II)	**2**-10;**15**-20,34;**27**-A5
Glossopalatine	**20**-III1	Olfactory	
Gluteus Maximus	**21**-69	Bulb (I)	**27**-A2
Gluteus Medius	**21**-66	Tract (I)	**27**-A3
Gracilis	**21**-75	with Intranasal Ramifications	**8**-D1
of Heart	**4**-A26,17,23	Peroneal, Common	**21**-99
Iliacus	**10**-D16	Pelvic	**30**-C5
Interossei, Dorsal	**21**-62	Phrenic	**7**-11;**8**-A11
Interventricular	**4**-A26	Left	**7**-18
Latissimus Dorsi	**21**-65	Right	**7**-21
Levator Palpebrae Superioris	**15**-24;**2**-3,31	Posterior Brachial Cutaneous	**21**-91
Mastication	**26**-38	Posterior Femoral Cutaneous	**21**-97
Muscular Tissue	**28**-B	Radial	**21**-89
Mylohyoid	**5**-8; **20**-IIB18	Deep Branch of Radial	**21**-92,93
Oblique, External	**21**-68	Sacral	**21**-86
Inferior (Eye)	**2**-15;**14**-14;**15**-22	Sciatic	**21**-98
Superior (Eye)	**2**-4;**14**-31;**15**-26	Sensory	
Opponens Digiti Minimi	**21**-63	of Skin	**3**-5
Orbicularis Oculi	**2**-20,32;**15**-36	with Dorsal Root Ganglion	**27**-C11
Papillary	**4**-A19	Spermatic Cord	**13**-A19,B14
Peroneus Longus	**21**-80	Splanchnic	**30**-C2,C3,C4
Psoas Major	**10**-D15	Thoracic	**21**-84
Rectus, of Eye		Tibial	**21**-100
Inferior	**2**-14;**14**-13	Trigeminal (V)	**8**-D6;**27**-A23
Lateral	**14**-9;**15**-23	Trochlear (IV)	**27**-A25
Medialis	**14**-29;**15**-25	Ulnar	**21**-94
Superior	**2**-5;**14**-1;**15**-24	Vagus	**8**-A1;**26**-21;**27**-A13;**30**-C1
Sartorius	**21**-76	to Larynx, Lung and Aorta, Heart	**8**-A1
Semimembranosus	**21**-73	Vestibular	**9**-24
Semitendinous	**21**-74	Vestibulocochlear (VII)	**27**-A21
Septum (Interventricular)	**4**-A26	Nervous Control of Respiration	
Soleus	**21**-79	and Symptoms of Infections	**8**
Sternocleidomastoid	**21**-54	Nervous System and Heart	**28**-C
of Stomach	**6**-D6	Neurons	**28**-C
Temporal	**15**-4	Types of	
Tendon Calcanean	**21**-82	Neutrophils	**17**-NBC
Tensor Fasciae Latae	**21**-67	Normal Quantity	
Tensor Tympani	**9**-A29	Mature	**16**-21
Teres Major	**21**-64	NODES: See Lymph Nodes	
Triceps	**21**-56	Normoblast	**16**-13;**18**-E15,E23,G11,I11
Vastus Lateralis	**21**-71	NOSE	
Vastus Medialis	**21**-77	Meatus	**8**-B4,B9
Zygomaticus Major	**26**-41	Nasal Bone	**1**-5;**14**-27;**15**-29;**19**-11
Muscular Tissue	**28**-B	Nasal Cartilage	**14**-18
Myeloblast	**16**-2	Nasal Cavity	**7**-35
MYELOCYTE		Cross Section of	**8**-B5
Basophilic	**16**-11;**18**-H14	Nasal Passage	**20**-IIB22
Eosinophilic	**16**-10;**18**-H15	Nasal Septum	**8**-B12;**15**-28
Neutrophilic	**16**-9;**18**-H12,H13,H18		
Myometrium	**11**-J15		
Myopia	**2**-B	**O**	
		Occipital Bone	**19**-123
		Occipital Lymph Nodes	**24**-A2
N		Occipital Protuberance	**21**-3
Nasal Concha	**8**-B10,B11	Olfactory Bulb	**27**-A2
Nasal Process	**15**-30	Olfactory Tract	**27**-A3
Nasopharynx	**7**-5;**8**-A14;**20**-IIB2	Optic Chiasma	**15**-33;**27**-A4,D25
Nasolacrimal Duct	**14**-17	Optic Papilla	**2**-9
Neck, Sagittal Section	**26**	Optic Tract	**15**-19;**27**-A7
Nephron	**10**-A4,C	Oral Cavity	**7**-32
NERVES		Oropharynx	**7**-7;**8**-A13;**18**-13;**20**-IIB6
Nerves	**21**	Ovary (Ovaries)	**11**-J19;**12**-A3;**22**-7;**23**-7b
Abducens (VI)	**27**-A11	Follicles	**11**-J3
Accessory (XI)	**27**-A15	Opened; Section	**11**-J8
Auditory	**9**-23		
Brachial Plexus	**21**-88		

Anatomical Feature	Chart number, bold Feature number, light
Lateral Duct	9-B12
Lead Poisoning	18-E6 to E12
Lens	15-37
LEUCOCYTE: see Blood Cells, White	
LEUKEMIA	
Acute	18-I
Lymphatic, Chronic	18-G
Myelocytic, Chronic	18-H
LIGAMENTS	
Arteriosum	4-A10,B13
Broad, of Uterus	11-J17
Cricothyroid	20-IIB15,III7
Palpebral	
Lateral	14-8
Medial	14-26
Round, of Uterus	11-J18;12-A6
Ovarian	11-J22
Suspensory, of Crystalline Lens	2-35
Sacrouterine	12-A19
LIMBS	
Lower	11-K4
Upper	11-K6
LIVER	
Liver	19-101;25-J17;30-E8
Diseased	6-A8
Healthy	5-42,46
Lymph Nodes and Vessels	25-J
LOBE	
Frontal, of Brain	26-53
Hypophysis	27-D27,D28
Lung	7-16,24
Occipital	26-13
Parietal	26-6
Temporal	26-14
Lumbar Vertebrae	1-24;21-18
LUNGS	21-102;30-E6
Alveolar Sacs (Magnified View)	7-14
Alveolus (Magnified View)	7-15
Left	4-A9;5-15;19-25;26-29
Apex of	7-12
Cross Section of	7-16
Right	4-A1;5-48;7-24;19-108
Vessels and Lymph Nodes	24-A7
LYMPHATIC SYSTEM	24;25
LYMPH NODES	
Aortic, Lateral	24-A23
Auricular, Posterior	24-A1
Axillary	24-A5,26
Buccal	24-A35
Capillaries	6-C2;24-C
Capsule	25-B5
Celiac	25-K10
Cervical	24-A3,A32
Superior, Deep	24-A32
Colic, Right	25-J12
Cross Section	16-6;24-B
Follicle with Germinal Center	24-B4
Gastric	
Superior	25-J4
Interior	25-J5
Hepatic	25-J19
Hypogastric	24-A14
Iliac	
Common	24-A21
External	24-A20
Inguinal	24-A16,A19
Magnified	24-B
Mandibular	24-A33
Maxillary	24-A36
Medullary Sinus	24-B7
Mesenteric	25-K6
Mesocolic	25-J7
Nodule	6-C5
Cortical	24-B3
Occipital	24-A2
Pancreaticolienal	25-K3,J2
Paracardial	25-J1
Parotid	24-37
Pleura and Lung	24-A7
Sacral	24-A15
Splenic	25-K2
Subclavicular	24-A28
Subinguinal	24-A18
Submental	24-A34
Subpyloric	25-K9

Anatomical Feature	Chart number, bold Feature number, light
LYMPH NODES (Continued)	
Superficial	24-A24
Supratrochlear	24-A22
Trabecula	24-B2
Tracheal	24-A31
Lymphoblast	16-7; 18-G12
LYMPHOCYTE	17-NBC; 18-A9,B12,C11,D13 F9,G13,I13
Atypical	18-F8
Large	16-25
Small	16-17
LYMPH VESSELS	24-F2; 25
Afferent	24-B1,E1
Chyliferous	25-G2
Cisterna Chyli	24-A12
Cross Section	24-F
Efferent	24-B6,E2
Ileocolic Chain	25-J9
Inguinal	24-A16
Lumbar Chain	24-A13
Mediastinal	24-A27
Paracolic Chain	25-J8
Parotid	24-A37
of Pleura and Lung	24-A7
Intercostal	24-A9
Renal	24-A11
of Skin	24-F
of Small Intestines	25-K
Supra Trochlear	
Splenic	24-A10
of Stomach, Liver, and Large Intestines	25-J
Subpyloric	25-J15
M	
Macrocyte	18-B6, E18
Male Reproductive Organs	13
Malleus	9-A3
Mamillary Body	27-A29,D29
Mandible	1-8;19-16
Ascending Ramus of	21-6
Manubrium	1-11; 19-23
Massa Intermedia	27-B8,D13
Mastoid Process	9-A7;21-5
MATTER	
Gray	27-C8
White	27-C7
Median Sagital Section of Brain	27
MEDULLA	
of Adrenal Gland	23-4a
of Hair	3-1
of Kidney	10-A3,
Oblongata	7-6;26-19;27-D31,A18;30-B2
Spinalis	7-8
Pyramids	10-D7
Inner Zone	10-C13
Outer Zone	10-C14
Megakaryocyte	16-12
Megaloblast	18-B10
MEMBRANE	
Development in Embryo	11-H
Mucous	
Inflammatory Hyperemia	8-D2
Inflammatory Vascular Dilitation	8-E1
Thyrohyoid	20-III5
Mesosalpinx	11-J21
Mesentery	6-B
Metacarpals	1-34
Metamyelocyte	
Basophilic	16-16
Eosinophilic	16-15;18-H17
Neutrophilic	16-14;18-H16
Metatarsals	1-44;19-77;21-52
Microcyte	18-C6,D6
Mid-Brain	15-15;26-15;30-B1
Modiolus	9-A20
Monoblast	16-3
Monocyte	16-24;17-NBC;18-A10,F7
Mononucleosis, Infectious	18-F
MOUTH	

Anatomical Feature	Chart number, bold Feature number, light	Anatomical Feature	Chart number, bold Feature number, light
BONES (Continued)		BRAIN (Continued)	
Nasal Process of Medial Epicondyle	15-30	Cortical Area of Larynx	26-50
1-37	Eyes, Conjugated Movement	26-55
Medial Malleolus	1-46	Fingers, Function of	26-54
Metacarpals	1-34;19-85;21-36	Functional Areas of Brain	
Metatarsals	1-44;19-77;21-52	Auditory Speech Area	26-11
Nasal	1-5;14-27;15-29;19-11	Ankle	26-59
Nasion	19-10	Conjugated Movements of Eyes	26-55
Navicular	19-75	Cortical Area of Larynx	26-50
Obelion	19-125	Fingers	26-54
Occipital	19-123;21-2	Hearing	26-10
Ophryon	19-7	Hip	26-58
Parietal	1-2;19-126;21-1	Motor Speech Area	26-48
Patella	1-39;19-82	Mouth	26-52
Phalanges	1-35,45;19-84,70;21-37,53	Muscle Sensation	26-5
Pisiform	19-48;21-35	Post Central Gyrus (Sensory Area)	26-2
Pterion	19-5	Shoulder	26-57
Pubis	19-56	Smell	26-45
Pubic Symphysis	1-31;12-A9;13-A7;19-89	Taste	26-46
Radius	1-27;19-95;21-27	Tongue	26-51
Ribs	21-17	Vision	26-12
Cross Section	16-4	Visual Speech Area	26-9
Eleventh Thoracic	10-D23	Wrist	26-56
First	26-33	Writing	26-7
Twelfth	1-20	Gray Matter	15-5;27-C8
Sacrum	1-28;21-19	Hearing, Function of	26-10
Scaphoid	19-40;21-28	Hip, Function of	26-58
Scapula	1-15;19-24;21-22;26-25	Horizontal Section	27-B
Spine of	21-21	Lobe	
Skull	15-6;26-4	Frontal	26-53
Sphenoid	1-3;19-9	Occipital	26-13
Stephanion	19-4	Parietal	26-6
Sternum		Temporal	26-14
Body of	1-16 19-29;26-30	Median Sagittal Section	27-D
Manubrium of	1-11	Mid-Brain	15-15;26-15;30-B1
Styloid Process	9-A8	Motor Speech Area	26-48
Talus	19-79	Mouth, Function of	26-52
Tarsals	1-43	Muscle Sensation Area	26-5
Temporal	1-4;9-A32;19-2;21-4	Pole	
Inferior Line	19-1	Frontal	27-D19,B5,A1
Mastoid process of	9-A7;21-5	Occipital	27-D39,B13
Petrous Part of	9-A25	Temporal	27-A8
Styloid Process of	9-A8	Pre-Central Gyrus (Motor Area)	26-1
Superior Line	19-127	Section, Horizontal	27-B
Tibia	1-41;19-81;21-45	Sensory Cortex	27-D8
Trapezium	19-41;21-31	Shoulder, Function of	26-57
Trapezoid	19-42;21-32	Sight as a function of Brain	15
Triquetrum	19-47;21-30	Smell, Function of	26-45
Trochanter		Sulcus, Central	26-3
Greater	19-59;21-43	Sylvian Fissure	26-49
Lesser	19-60;21-44	Taste, Function of	26-46
Ulna	1-26;19-96;21-26	Tela Choroidea of Third Ventricle	27-D12
Vertebra		Tongue, Function of	26-51
Cervical	1-9;21-15	Uncus	27-A27
Lumbar	21-18	Ventricle, Fourth	15-10;27-D35
Fifth Lumbar	1-24	Lateral, Anterior Horn	27-B2
Thoracic	21-16	Inferior Horn	27-B19
Twelfth Thoracic	1-19	Posterior Horn	27-B15
Vertebral Column	26-23	Third	27-D10,B21
Vertex	19-128;21-9	Vision, Function of	26-12
Xiphoid Process	1-18	Visual Speech Area	26-9
Zygomatic Arch	1-6;14-10;26-43	White Matter	27-C7
Zygomatic (Malar)	15-1	Wrist, Function of	26-56
Bowman's Capsule	10-C1B	Writing, Function of	26-7
Brachial Plexus	21-88	Branchial Arches	11-K2
BRAIN		Breathing, Position of Larynx and Pharynx in	20-IIB
Brain	15-7; 26;27	Bregma	19-3
Ankle, Function of	26-59	Bronchioles	7-17
Anterior Commissure	27-D32	Bronchitis, Acute Tracheo	8-E
Auditory Speech Area	26-11	BRONCHUS (BRONCHI)	
Base of Brain	27-A	Inflamed, Cross Section	8-E
Anterior Perforated Surface	27-A30	Left Main	5-14;7-13
Callosal Sulcus	27-D16	Ramification	7-17
Caudate Nucleus	27-B6		
Centrum Semiovale (White Matter)	15-7	**C**	
Cerebellum	26-17;27-A-16;27-D36	Calcaneus	19-71;21-47
Cerebrum	21-101	Calcar Avis	27-B16
Cingulate Gyrus	27-D7,D17	Candida Albicans	20-IVB
Cingulate Sulcus	27-D14	Capitate Bone	19-46;21-33
Cortex of	2-2	CANAL	
		Central, of Spinal Cord	27-C6,D32

Colorprint®-Schick Anatomy Atlas
Index

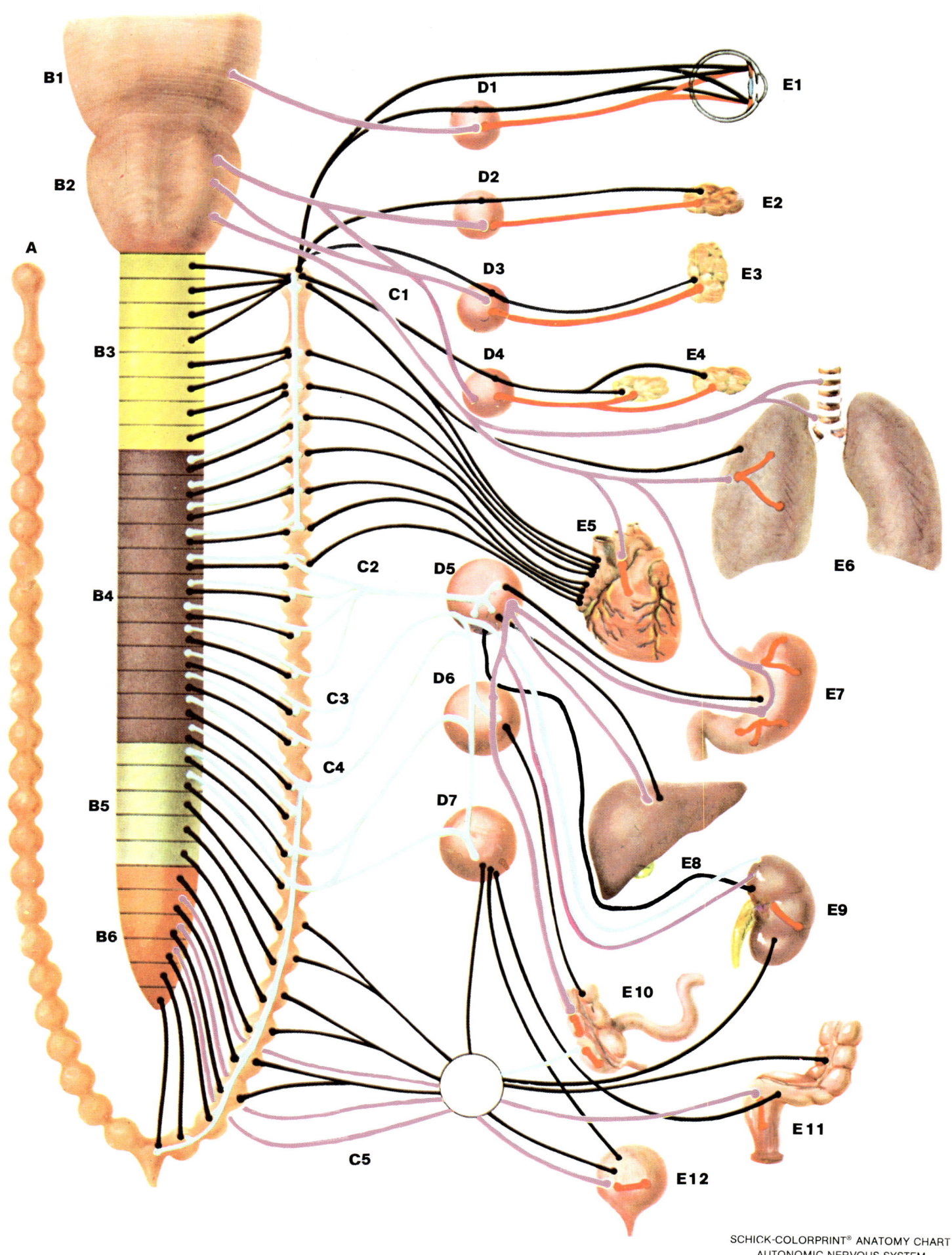

A

B1

B2

B3

B4

B5

B6

C1

C2

C3

C4

C5

D1

D2

D3

D4

D5

D6

D7

E1

E2

E3

E4

E5

E6

E7

E8

E9

E 10

E 11

E12

SCHICK-COLORPRINT® ANATOMY CHART
AUTONOMIC NERVOUS SYSTEM
No. NS30 © 1988 AMERICAN MAP CORP

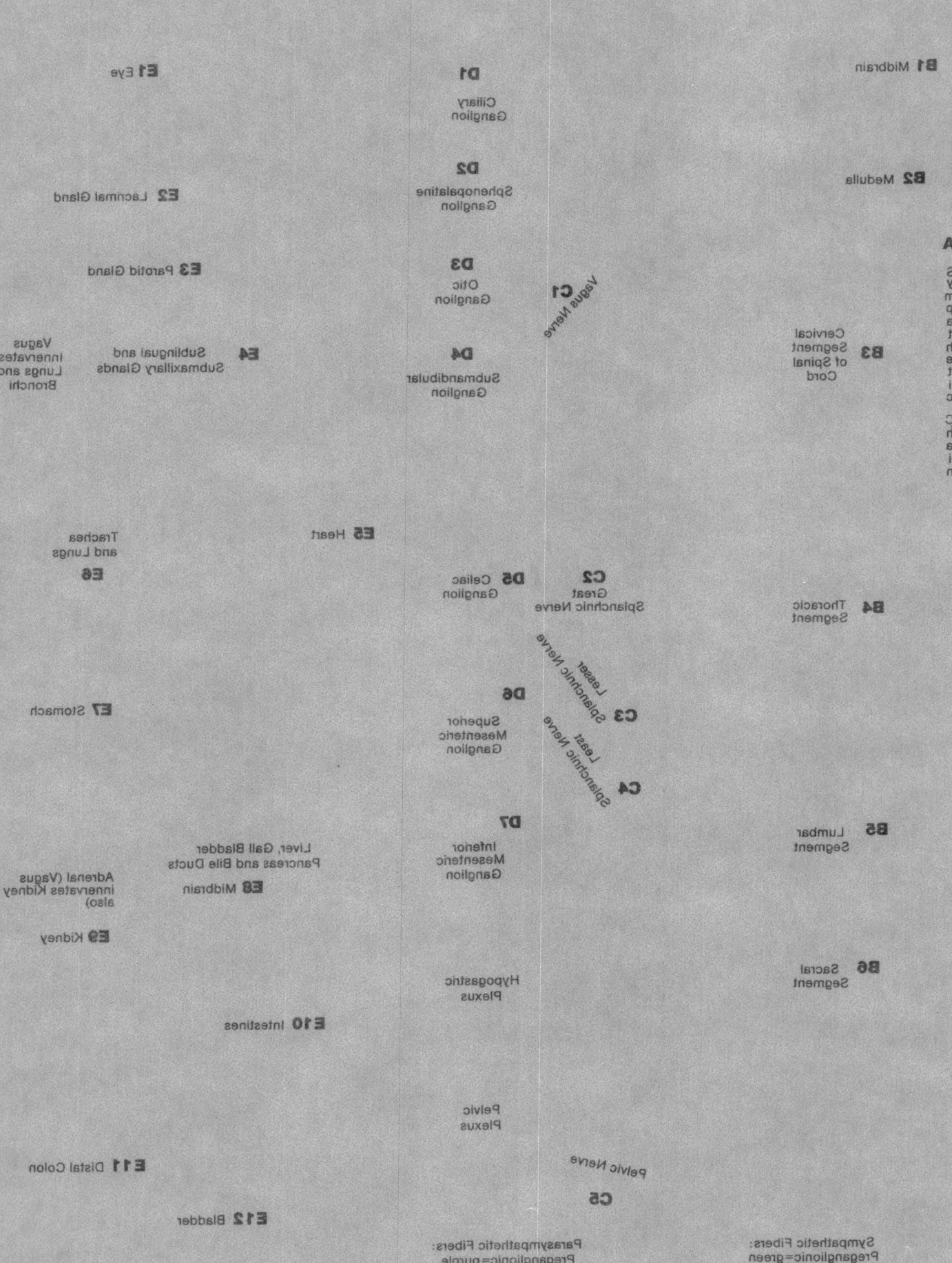

AUTONOMIC NERVOUS SYSTEM

B1 Midbrain

B2 Medulla

A Sympathetic Chain

B3 Cervical Segment of Spinal Cord

B4 Thoracic Segment

B5 Lumbar Segment

B6 Sacral Segment

D1 Ciliary Ganglion

D2 Sphenopalatine Ganglion

Vagus Nerve **C1**

D3 Otic Ganglion

D4 Submandibular Ganglion

E5 Heart

C2 Great Splanchnic Nerve

D5 Celiac Ganglion

Lesser Splanchnic Nerve

C3 Least Splanchnic Nerve

D6 Superior Mesenteric Ganglion

C4

D7 Inferior Mesenteric Ganglion

Hypogastric Plexus

Pelvic Plexus

Pelvic Nerve

C5

E1 Eye

E2 Lacrimal Gland

E3 Parotid Gland

E4 Sublingual and Submaxillary Glands

Vagus Innervates Lungs and Bronchi

Trachea and Lungs

E6

E7 Stomach

Liver, Gall Bladder Pancreas and Bile Ducts

E8 Midbrain

Adrenal (Vagus innervates Kidney also)

E9 Kidney

E10 Intestines

E11 Distal Colon

E12 Bladder

Sympathetic Fibers:
Preganglionic=green
Postganglionic=black

Parasympathetic Fibers:
Preganglionic=purple
Postganglionic=red

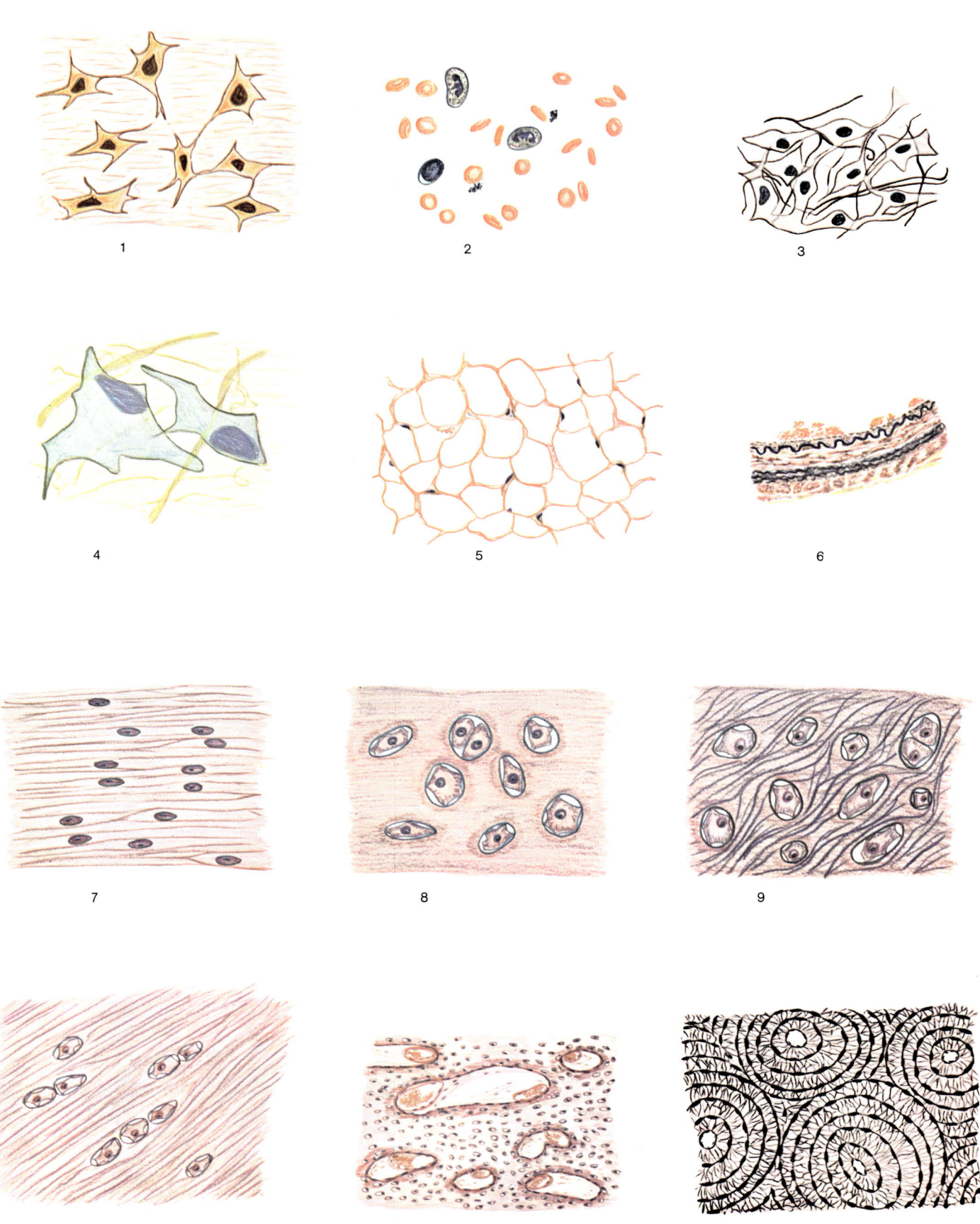

1

2

3

4

5

6

7

8

9

10

11

12

SCHICK-COLORPRINT® ANATOMY CHART
TISSUES
No.NS29 ©1988 AMERICAN MAP CORP

Connective Tissue

1 Mesenchymal

2 Blood

3 Reticular

4 Areolar (Loose)

5 Adipose

6 Elastic

7 Fibrous (Dense)

8 Hyaline Cartilage

9 Elastic Cartilage

10 Fibrocartilage

11 Cancellous Bone

12 Compact Bone

A

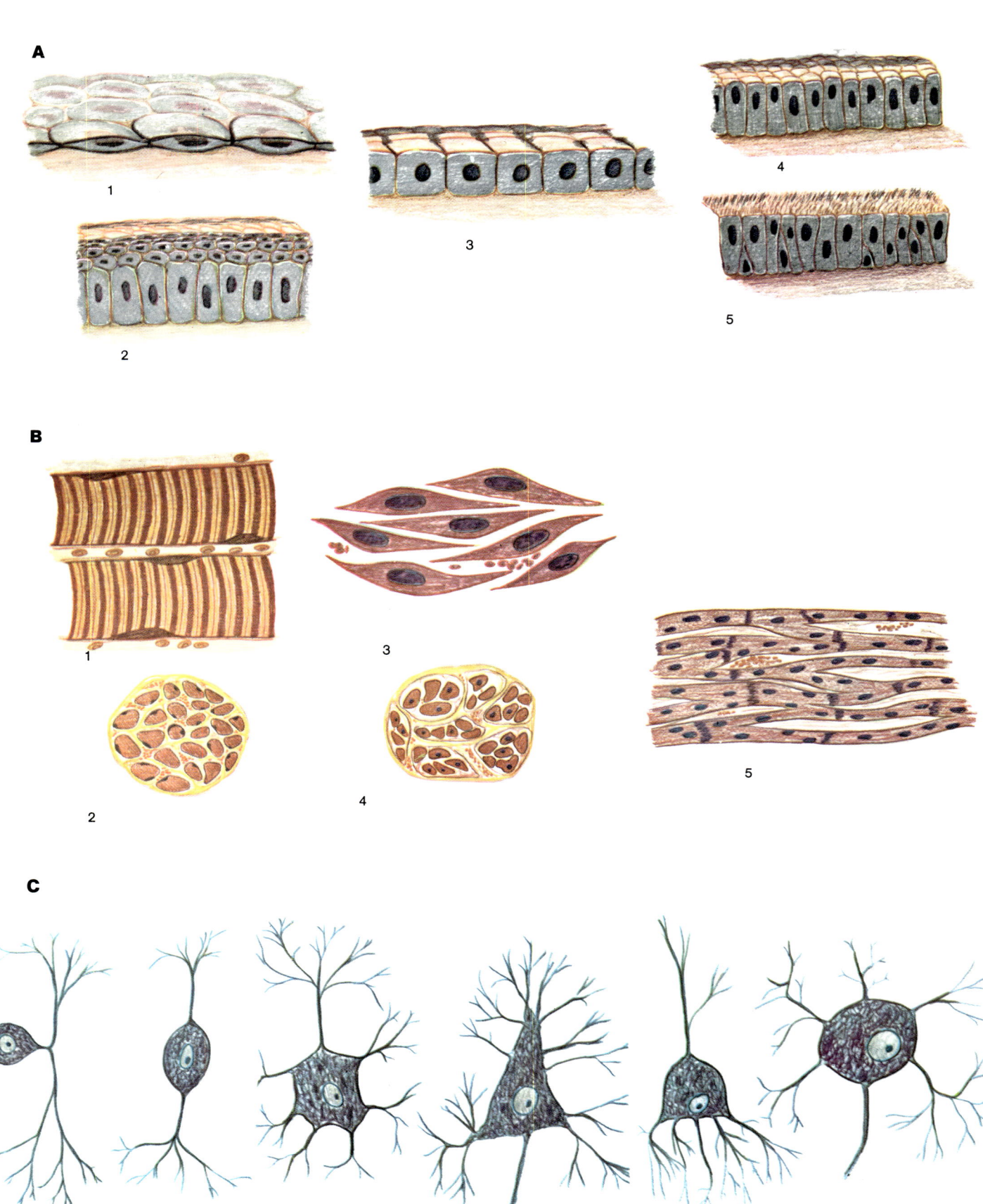

1

2

3

4

5

B

1

2

3

4

5

C

1   2   3   4   5   6

SCHICK-COLORPRINT® ANATOMY CHART
TISSUES
No. NS28 ©1988 AMERICAN MAP CORP

A. Epithelial Tissue

4 Simple Columnar

1 Simple Squamous

3 Cuboidal

5 Pseudostratified Columnar Ciliated

2 Stratified Squamous

B. Muscular Tissue

1 Skeletal Muscle Cells

3 Smooth Muscle Cells

5 Cardiac Muscle Cells

4 Bundles of Smooth Muscles

2 Bundles of Skeletal Muscles

C. Nervous Tissue

Various Types of Neurons

1 Pseudounipolar   2 Bipolar   3 Multipolar   4 Pyramidal Cell   5 Purkinje Cell   6 Autonomic Ganglion Cell

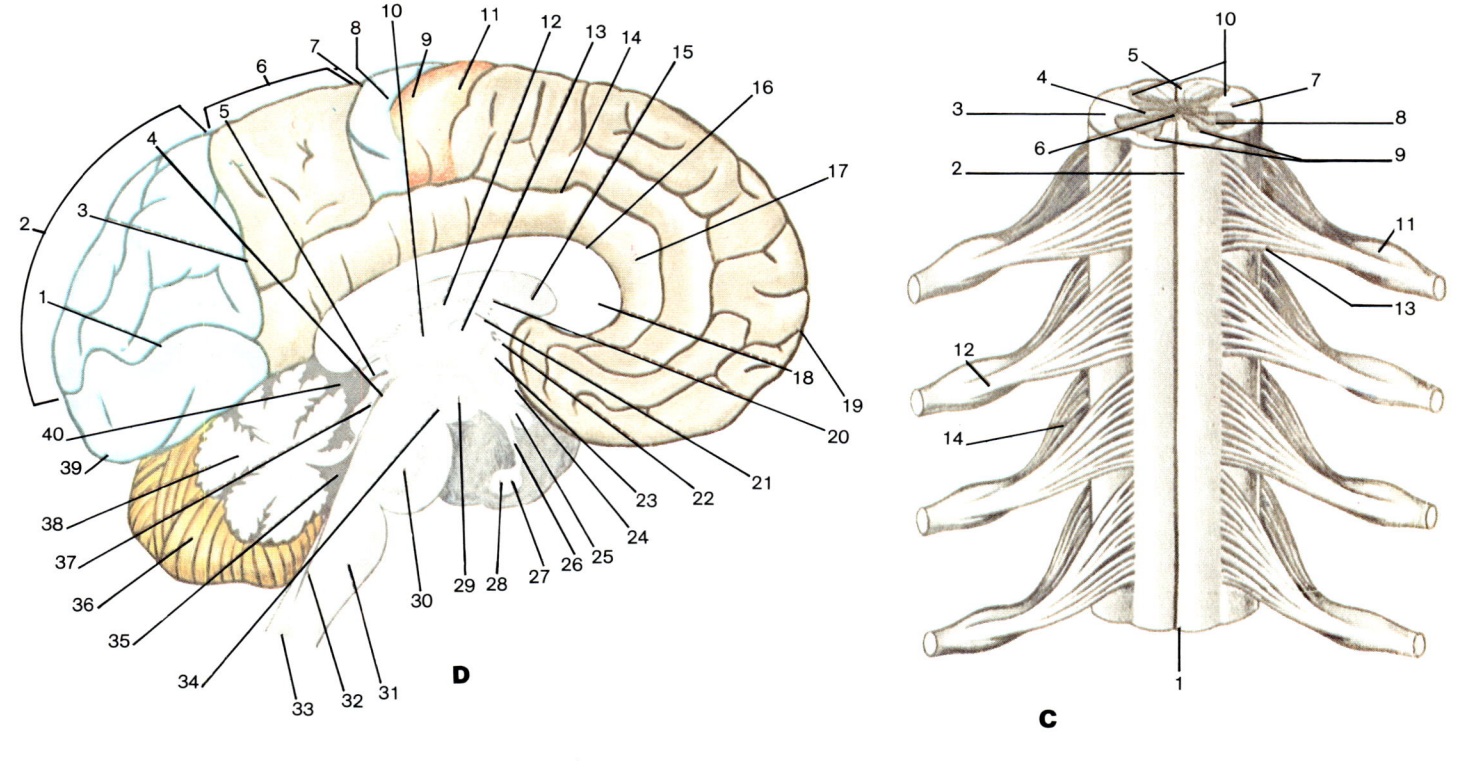

D

C

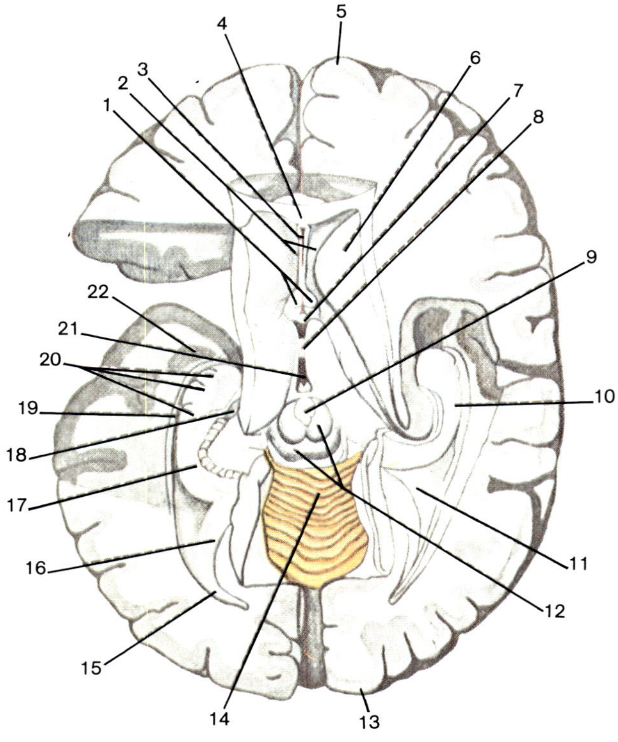

B

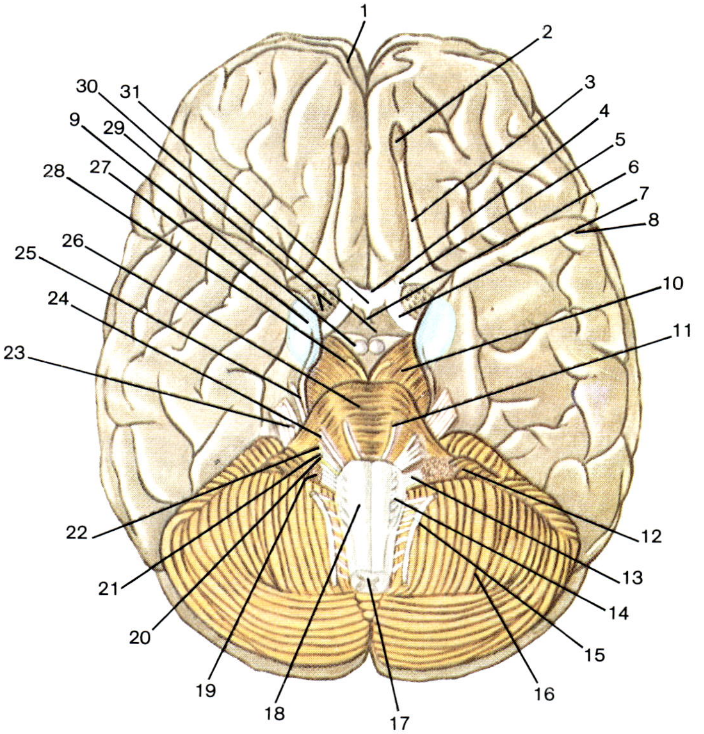

A

BRAIN
Median sagittal and horizontal sections, spinal cord, base of brain

D. Median Sagittal Section

1. Calcarine Sulcus
2. Cuneus
3. Parieto-Occipital Sulcus
4. Aqueduct of Sylvius (Cerebral Aqueduct)
5. Pineal Body
6. Precuneus
7. Cingulate Gyrus (Marginal Branch)
8. Sensory Cortex
9. Central Sulcus (of Rolando)
10. Third Ventricle
11. Motor Cortex
12. Tela Choroidea of Third Ventricle
13. Massa Intermedia (Interthalamic Adhesion)
14. Cingulate Sulcus
15. Septum Pellucidum
16. Callosal Sulcus
17. Cingulate Gyrus
18. Corpus Callosum
19. Frontal Pole
20. Fornix
21. Thalamus
22. Interventricular Foramen
23. Anterior Commissure
24. Lamina Terminalis
25. Optic Chiasma
26. Infundibulum
27. Anterior Lobe of Hypophysis (Pituitary)
28. Posterior Lobe of Hypophysis (Pituitary-Neurohypophysis)
29. Mammillary Body
30. Pons
31. Medulla Oblongata
32. Central Canal
33. Spinal Cord
34. Oculomotor Nerve
35. Fourth Ventricle
36. Cerebellum
37. Superior and Inferior Colliculi
38. Vermis
39. Occipital Pole
40. Cisterna Superior

C. Part of Spinal Cord (Frontal View, Dura Removed)

12. Superior and Inferior Colliculi
13. Occipital Pole
14. Vermis of Cerebellum
15. Posterior Horn of Lateral Ventricle
16. Calcar Avis
17. Fimbria of Hippocampus
18. Uncus of Hippocampal Gyrus
19. Inferior Horn of Lateral Ventricle
20. Hippocampal Digitations
21. Third Ventricle
22. Amygdaloid Complex

A. Base of Brain

1. Frontal Pole
2. Olfactory Bulb (I)
3. Olfactory Tract (I)
4. Optic Chiasma
5. Optic Nerve
6. Infundibulum
7. Optic Tract
8. Temporal Pole
9. Tuber Cinereum
10. Cerebral Peduncle
11. Abducens Nerve (VI)
12. Flocculus
13. Vagus Nerve (X)
14. Hypoglossal Nerve (XII)
15. Accessory Nerve (XI)
16. Cerebellum
17. Spinal Cord
18. Medulla Oblongata
19. Choroid Plexus
20. Glossopharyngeal Nerve (IX)
21. Vestibulocochlear Nerve (VII)
22. Intermediate Nerve
23. Trigeminal Nerve (V)
24. Facial Nerve (VII)
25. Trochlear Nerve (IV)
26. Pons
27. Uncus
28. Oculomotor Nerve (III)
29. Mammillary Body
30. Anterior Perforated Substance
31. Hypophysis (Pituitary)

No. NS27 © 1988 AMERICAN MAP CORP.

BRAIN
Median sagittal and horizontal sections, spinal cord, base of brain

1. Calcarine Sulcus
2. Cuneus
3. Parieto-Occipital Sulcus
4. Aqueduct of Sylvius
   (Cerebral Aqueduct)
5. Pineal Body
6. Precuneus
7. Cingulate Gyrus
   (Marginal Branch)
8. Sensory Cortex
9. Central Sulcus (of Rolando)
10. Third Ventricle
11. Motor Cortex
12. Tela Choroidea of Third Ventricle
13. Massa Intermedia
    (Interthalamic Adhesion)
14. Cingulate Sulcus
15. Septum Pellucidum
16. Callosal Sulcus
17. Cingulate Gyrus
18. Corpus Callosum
19. Frontal Pole
20. Fornix
21. Thalamus
22. Interventricular Foramen
23. Anterior Commissure
24. Lamina Terminalis
25. Optic Chiasma
26. Infundibulum
27. Anterior Lobe of Hypophysis
    (Pituitary)
28. Posterior Lobe of Hypophysis
    (Pituitary-Neurohypophysis)
29. Mammillary Body
30. Pons
31. Medulla Oblongata
32. Central Canal
33. Spinal Cord
34. Oculomotor Nerve
35. Fourth Ventricle
36. Cerebellum
37. Superior and Inferior Colliculi
38. Vermis
39. Occipital Pole
40. Cisternae Superior

D. Median Sagittal Section

1. Ventral Median Fissure
2. Ventral Funiculus
3. Lateral Funiculus
4. Dorsal Funiculus
5. Dorsal Median Sulcus
6. Central Canal
7. White Matter
8. Gray Matter
9. Ventral Horns
10. Dorsal Horns
11. Dorsal Root Ganglion
12. Motor Nerve
13. Ventral Root
14. Dorsal Root

**C. Part of Spinal Cord
(Frontal View, Dura Removed)**

1. Columns of Fornix
2. Anterior Horn of Lateral Ventricle
3. Septum Pellucidum
4. Corpus Callosum
5. Frontal Pole
6. Caudate Nucleus
7. Anterior Commissure
8. Massa Intermedia
   (Interthalamic Adhesion)
9. Pineal Body
10. Hippocampus
11. Collateral Eminence
12. Superior and Inferior Colliculi
13. Occipital Pole
14. Vermis of Cerebellum
15. Posterior Horn of Lateral Ventricle
16. Calcar Avis
17. Fimbria of Hippocampus
18. Uncus of Hippocampal Gyrus
19. Inferior Horn of Lateral Ventricle
20. Hippocampal Digitations
21. Third Ventricle
22. Amygdaloid Complex

1. Frontal Pole
2. Olfactory Bulb (I)
3. Olfactory Tract (I)
4. Optic Chiasma
5. Optic Nerve
6. Infundibulum
7. Optic Tract
8. Temporal Pole
9. Tuber Cinereum
10. Cerebral Peduncle
11. Abducens Nerve (VI)
12. Flocculus
13. Vagus Nerve (X)
14. Hypoglossal Nerve (XII)
15. Accessory Nerve (XI)
16. Cerebellum
17. Spinal Cord
18. Medulla Oblongata
19. Choroid Plexus
20. Glossopharyngeal Nerve (IX)
21. Vestibulocochlear Nerve (VII)
22. Intermediate Nerve
23. Trigeminal Nerve (V)
24. Facial Nerve (VII)
25. Trochlear Nerve (IV)
26. Pons
27. Uncus
28. Oculomotor Nerve (III)
29. Mammillary Body
30. Anterior Perforated Substance
31. Hypophysis (Pituitary)

A. Base of Brain

B. Horizontal Section

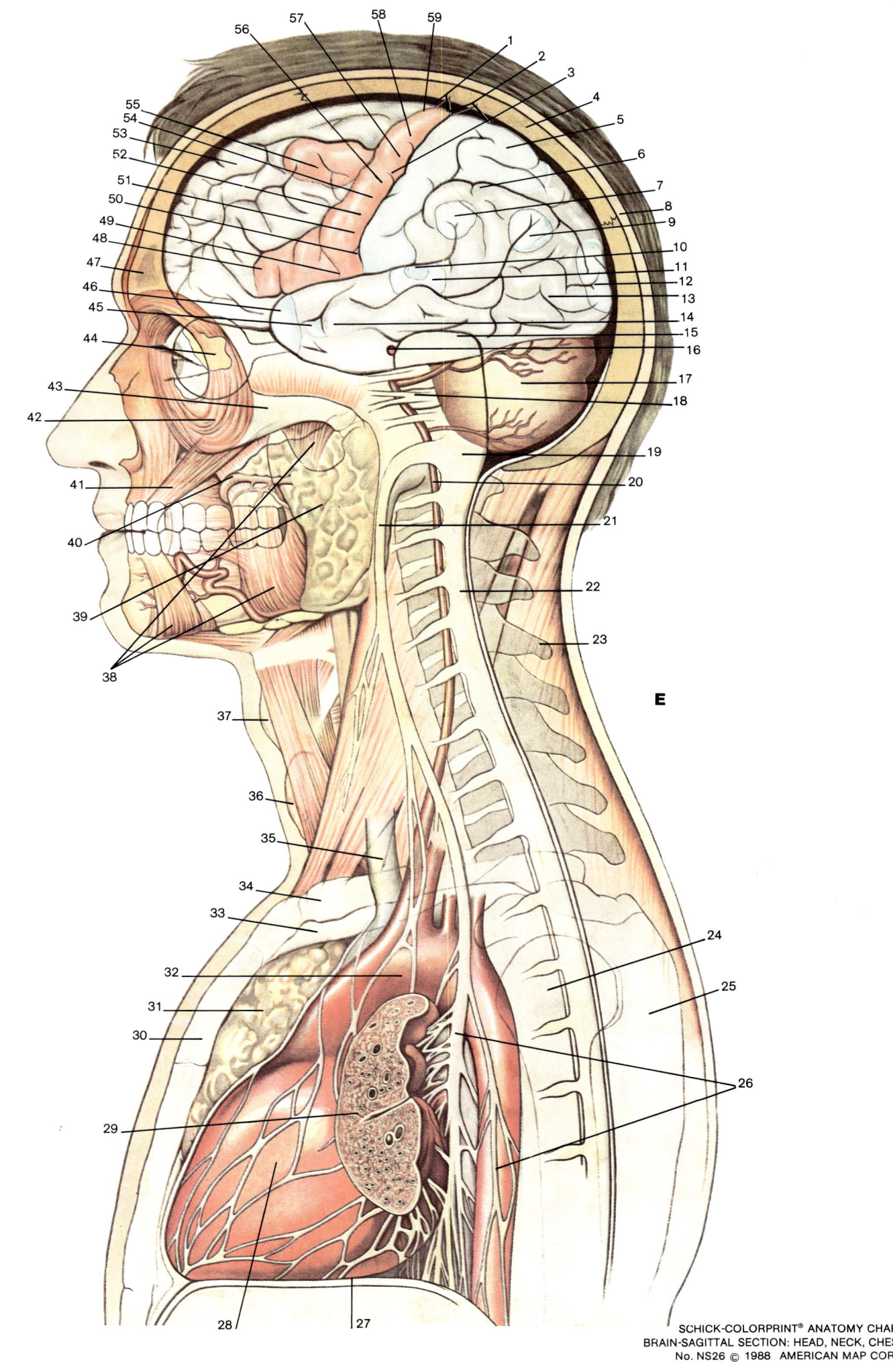

E

SCHICK-COLORPRINT® ANATOMY CHART
BRAIN-SAGITTAL SECTION: HEAD, NECK, CHEST
No. NS26 © 1988 AMERICAN MAP CORP.

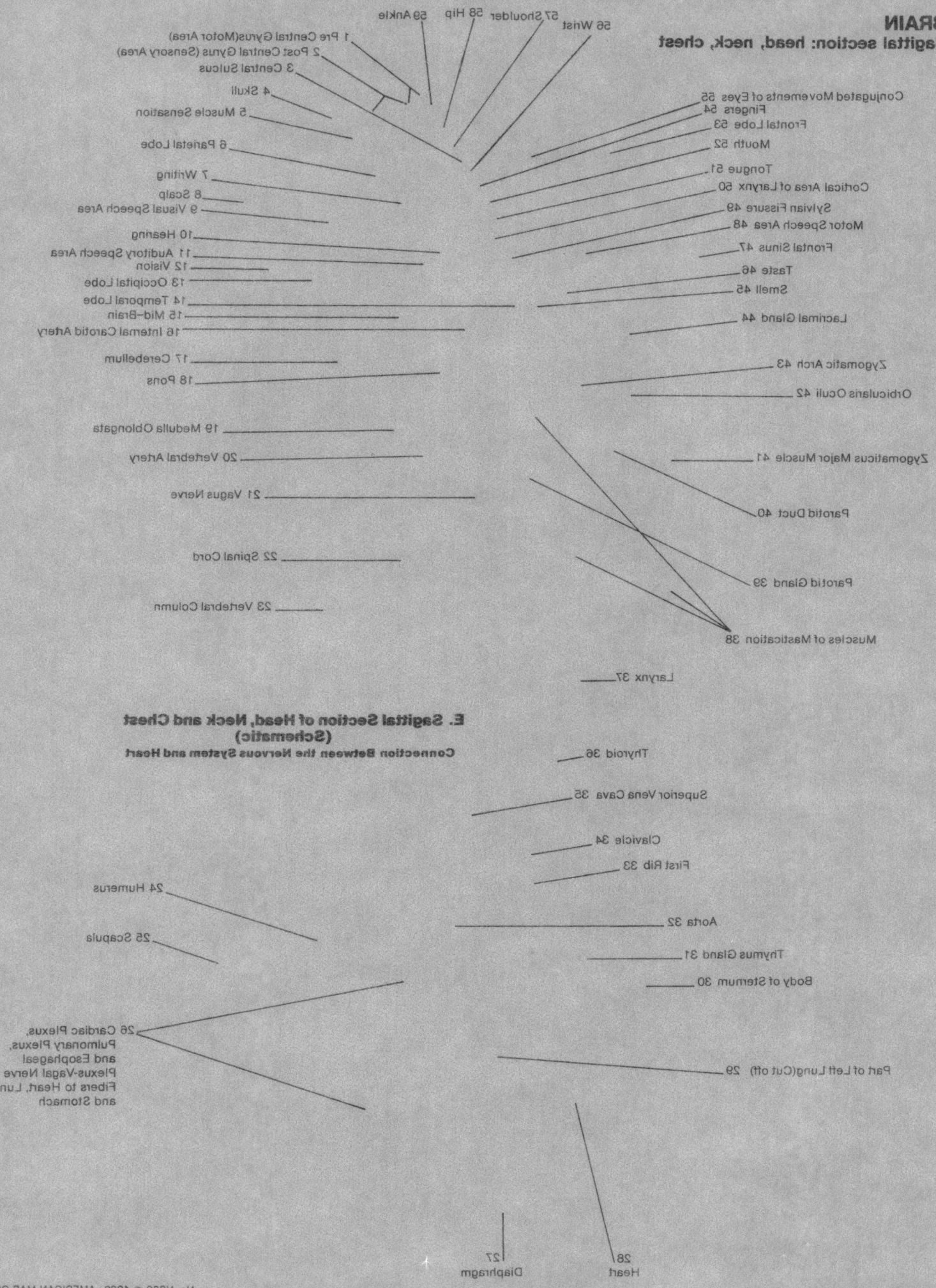

BRAIN
Sagittal section: head, neck, chest

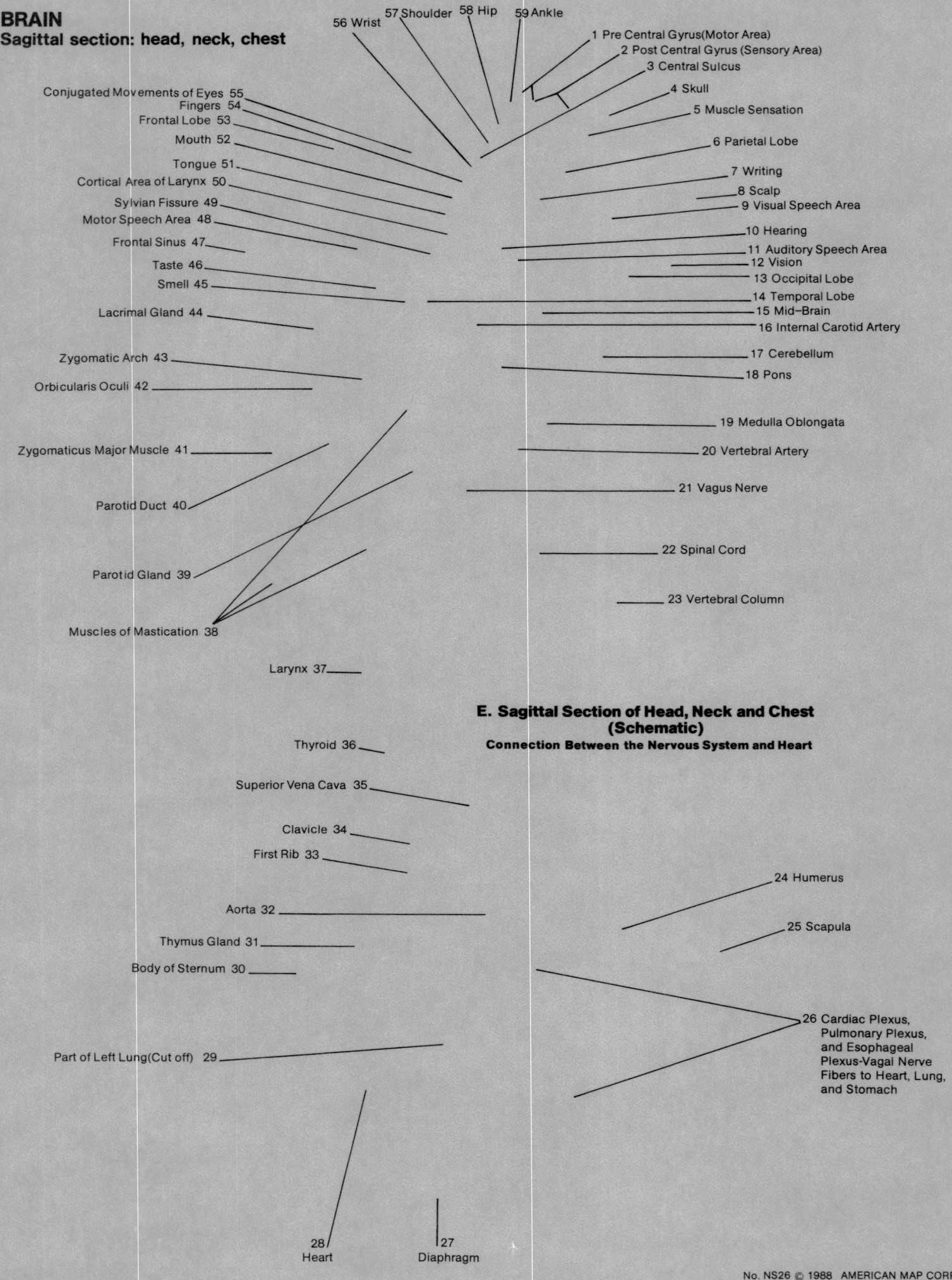

56 Wrist   57 Shoulder   58 Hip   59 Ankle

1 Pre Central Gyrus(Motor Area)
2 Post Central Gyrus (Sensory Area)
3 Central Sulcus
4 Skull
5 Muscle Sensation
6 Parietal Lobe
7 Writing
8 Scalp
9 Visual Speech Area
10 Hearing
11 Auditory Speech Area
12 Vision
13 Occipital Lobe
14 Temporal Lobe
15 Mid–Brain
16 Internal Carotid Artery
17 Cerebellum
18 Pons
19 Medulla Oblongata
20 Vertebral Artery
21 Vagus Nerve
22 Spinal Cord
23 Vertebral Column

Conjugated Movements of Eyes 55
Fingers 54
Frontal Lobe 53
Mouth 52
Tongue 51
Cortical Area of Larynx 50
Sylvian Fissure 49
Motor Speech Area 48
Frontal Sinus 47
Taste 46
Smell 45
Lacrimal Gland 44
Zygomatic Arch 43
Orbicularis Oculi 42
Zygomaticus Major Muscle 41
Parotid Duct 40
Parotid Gland 39
Muscles of Mastication 38
Larynx 37
Thyroid 36
Superior Vena Cava 35
Clavicle 34
First Rib 33
Aorta 32
Thymus Gland 31
Body of Sternum 30
Part of Left Lung(Cut off) 29

E. Sagittal Section of Head, Neck and Chest (Schematic)
Connection Between the Nervous System and Heart

24 Humerus
25 Scapula
26 Cardiac Plexus, Pulmonary Plexus, and Esophageal Plexus-Vagal Nerve Fibers to Heart, Lung, and Stomach

28 Heart
27 Diaphragm

No. NS26 © 1988 AMERICAN MAP CORP.

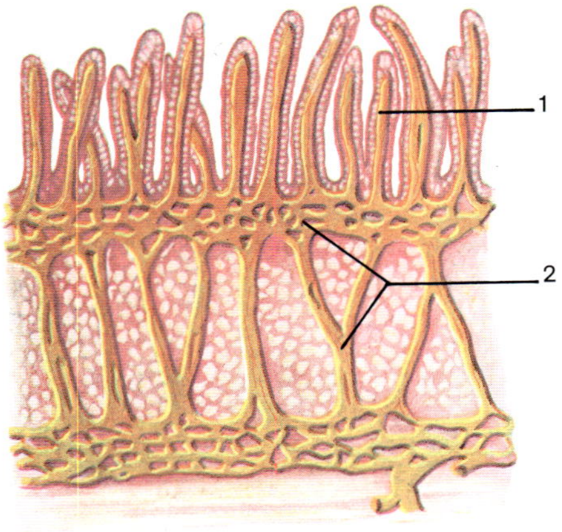

G

1

2

I

1

2

H

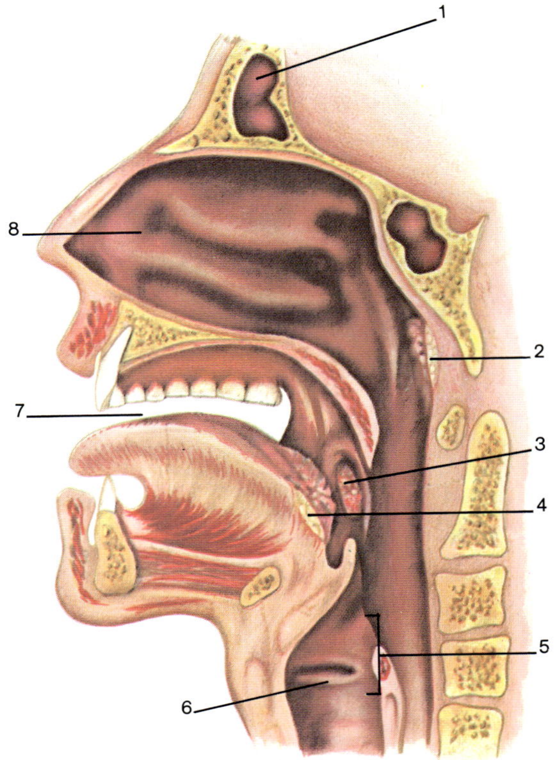

1

8

2

7

3

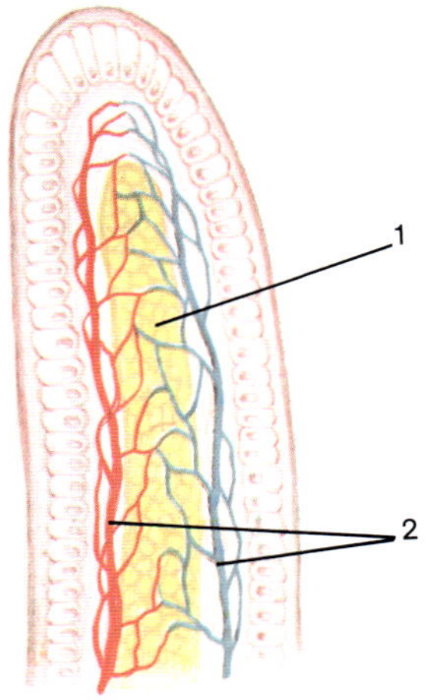

4

6

5

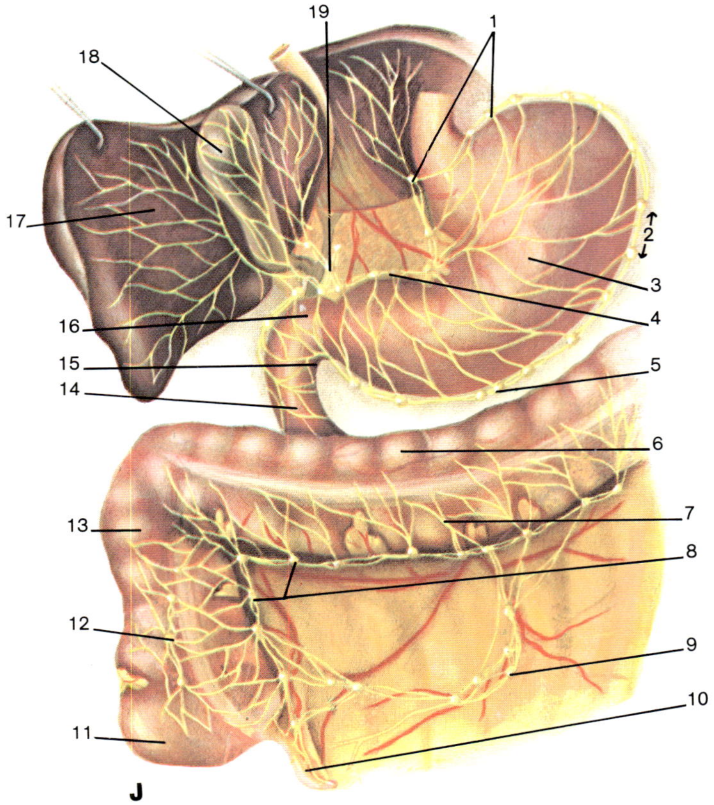

19

18

1

17

2

3

16

4

15

5

14

6

13

7

8

12

9

11

10

J

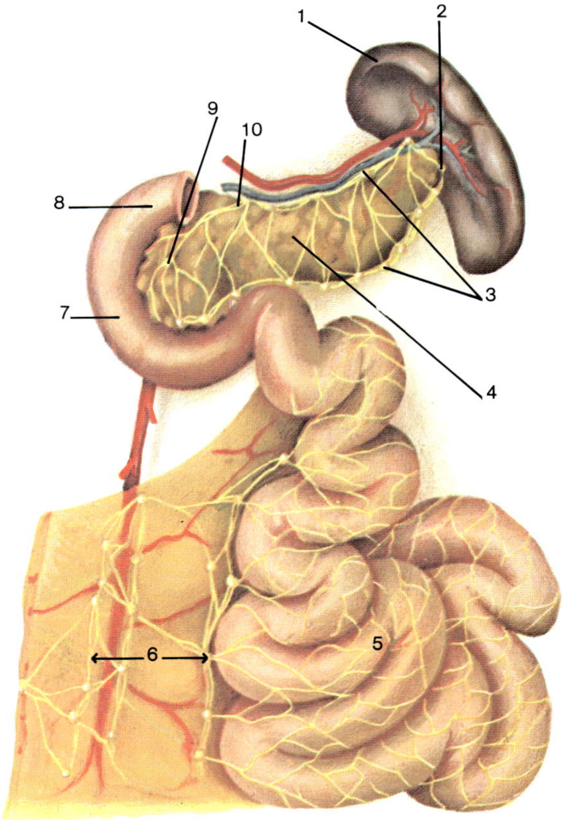

1

2

9

10

8

3

7

4

6

5

K

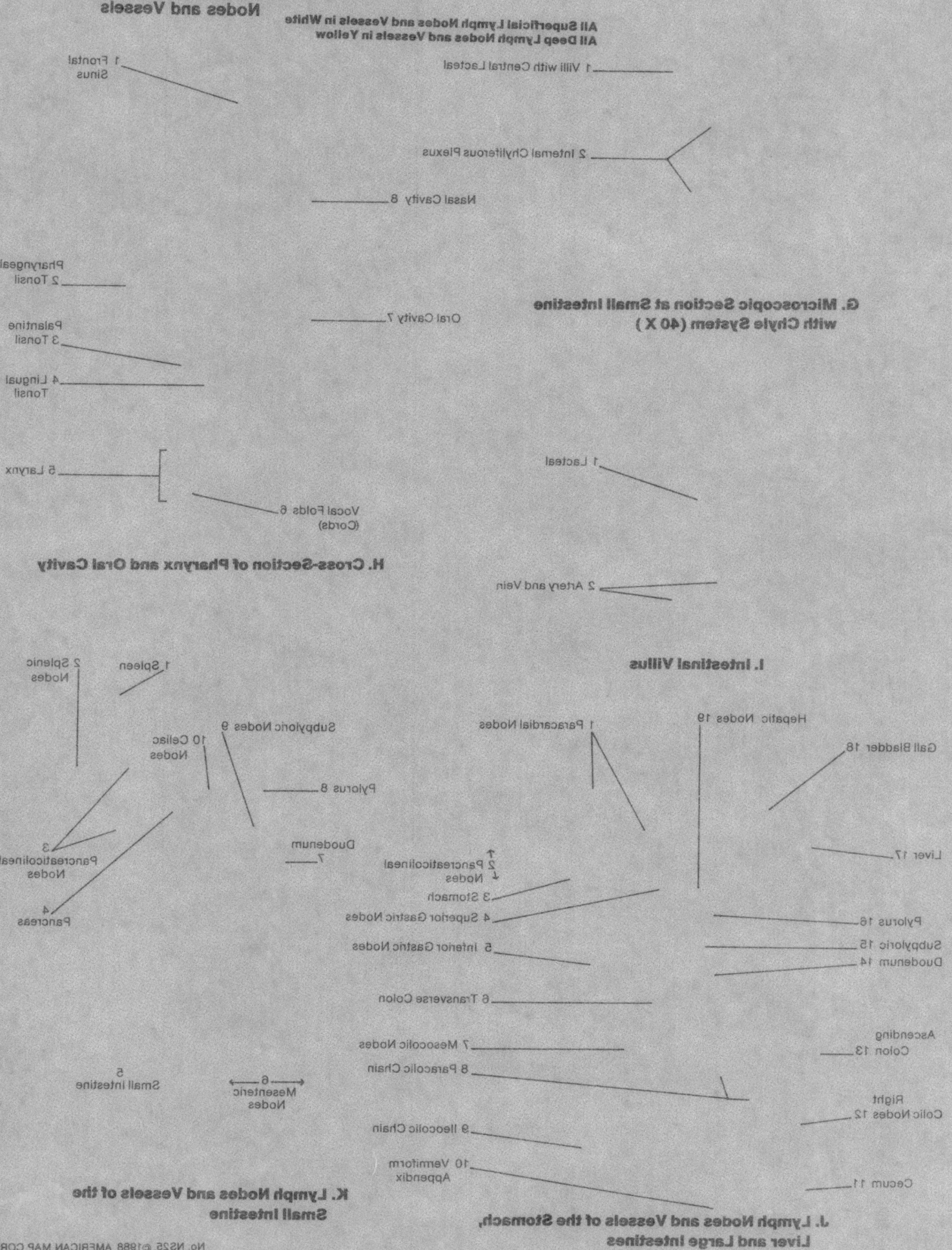

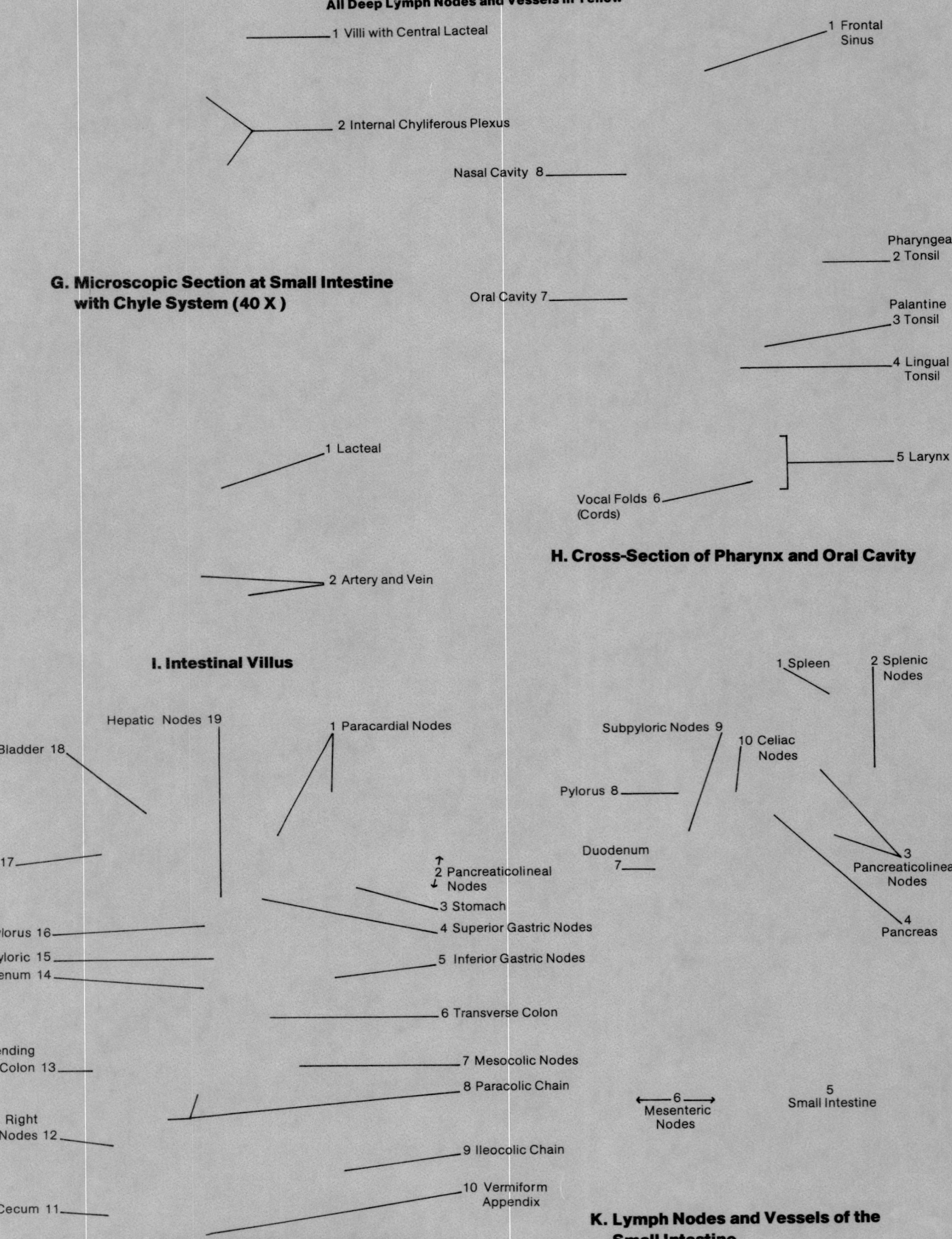

LYMPHATIC SYSTEM
Nodes and Vessels

All Superficial Lymph Nodes and Vessels in White
All Deep Lymph Nodes and Vessels in Yellow

1 Villi with Central Lacteal

1 Frontal Sinus

2 Internal Chyliferous Plexus

Nasal Cavity 8

G. Microscopic Section at Small Intestine with Chyle System (40 X)

Pharyngeal 2 Tonsil

Oral Cavity 7

Palantine 3 Tonsil

4 Lingual Tonsil

1 Lacteal

5 Larynx

Vocal Folds 6 (Cords)

2 Artery and Vein

H. Cross-Section of Pharynx and Oral Cavity

I. Intestinal Villus

1 Spleen   2 Splenic Nodes

Hepatic Nodes 19

1 Paracardial Nodes

Subpyloric Nodes 9   10 Celiac Nodes

Gall Bladder 18

Pylorus 8

Liver 17

Duodenum 7

3 Pancreaticolineal Nodes

↑ 2 Pancreaticolineal ↓ Nodes

4 Pancreas

3 Stomach

Pylorus 16

4 Superior Gastric Nodes

Subpyloric 15

Duodenum 14

5 Inferior Gastric Nodes

6 Transverse Colon

Ascending Colon 13

7 Mesocolic Nodes

8 Paracolic Chain

← 6 → Mesenteric Nodes

5 Small Intestine

Right Colic Nodes 12

9 Ileocolic Chain

10 Vermiform Appendix

Cecum 11

K. Lymph Nodes and Vessels of the Small Intestine

J. Lymph Nodes and Vessels of the Stomach, Liver and Large Intestines

No. NS25 ©1988 AMERICAN MAP CORP.

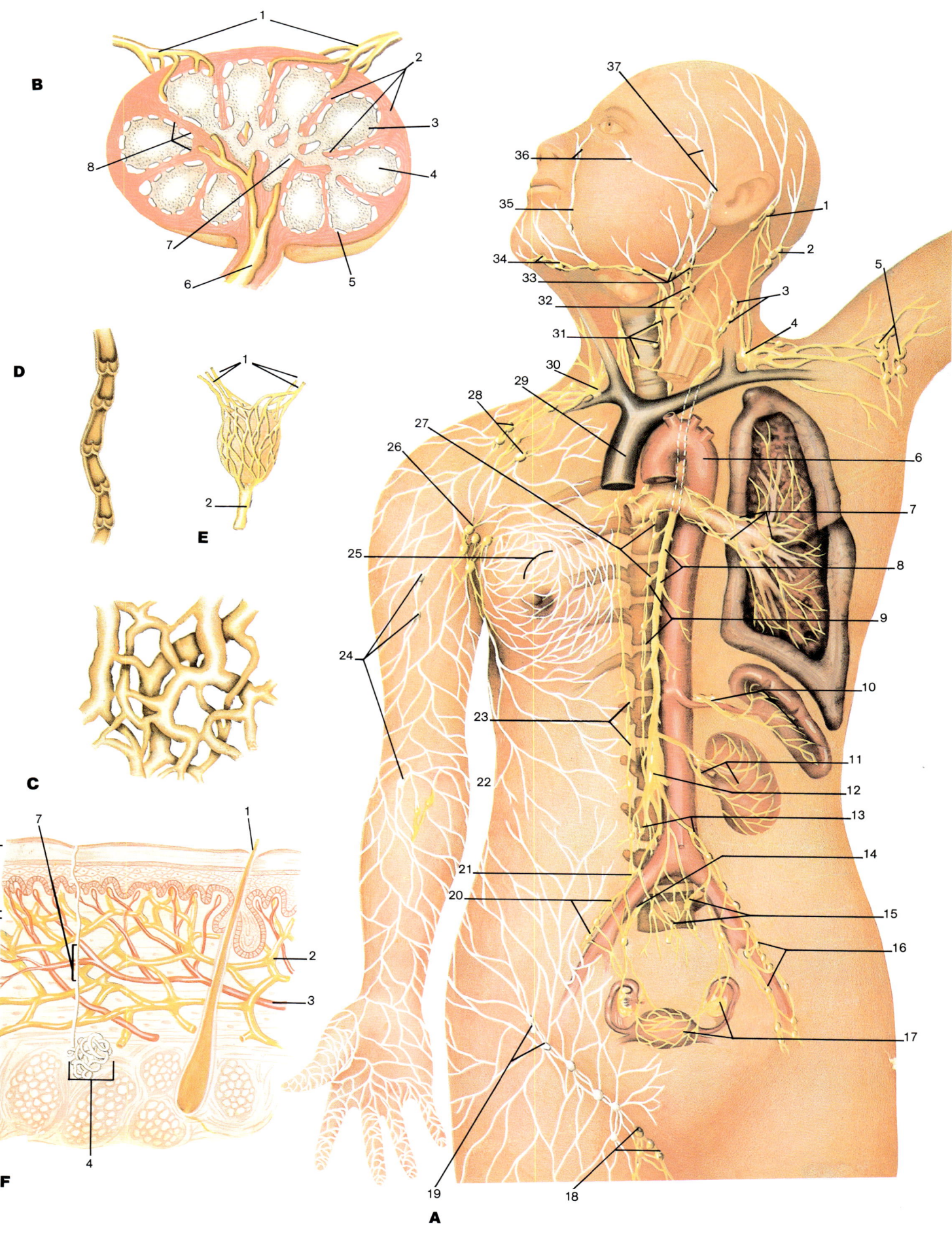

SCHICK-COLORPRINT® ANATOMY CHART
LYMPHATIC SYSTEM-GENERAL
NO. NS24 © 1988 AMERICAN MAP CORP.

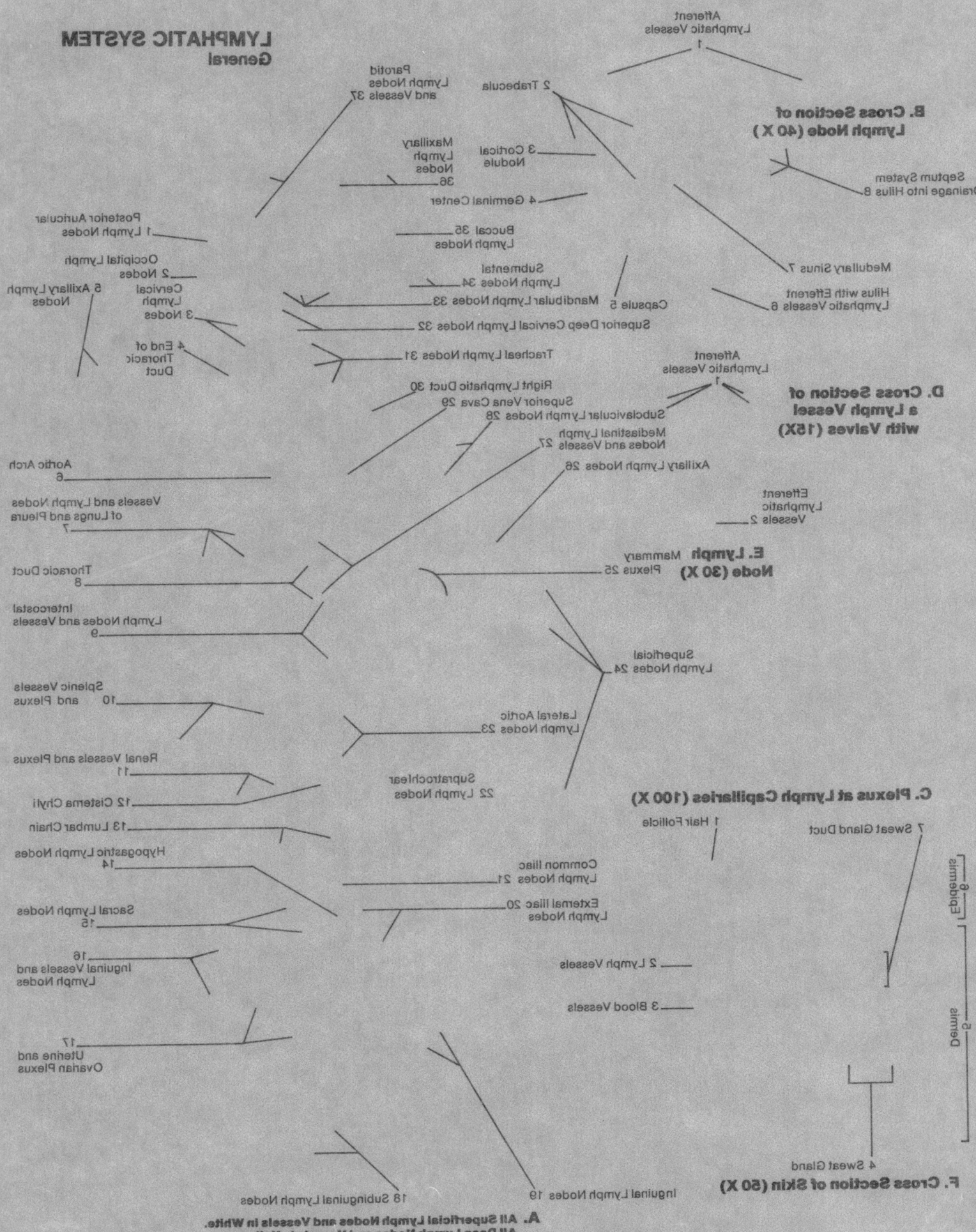

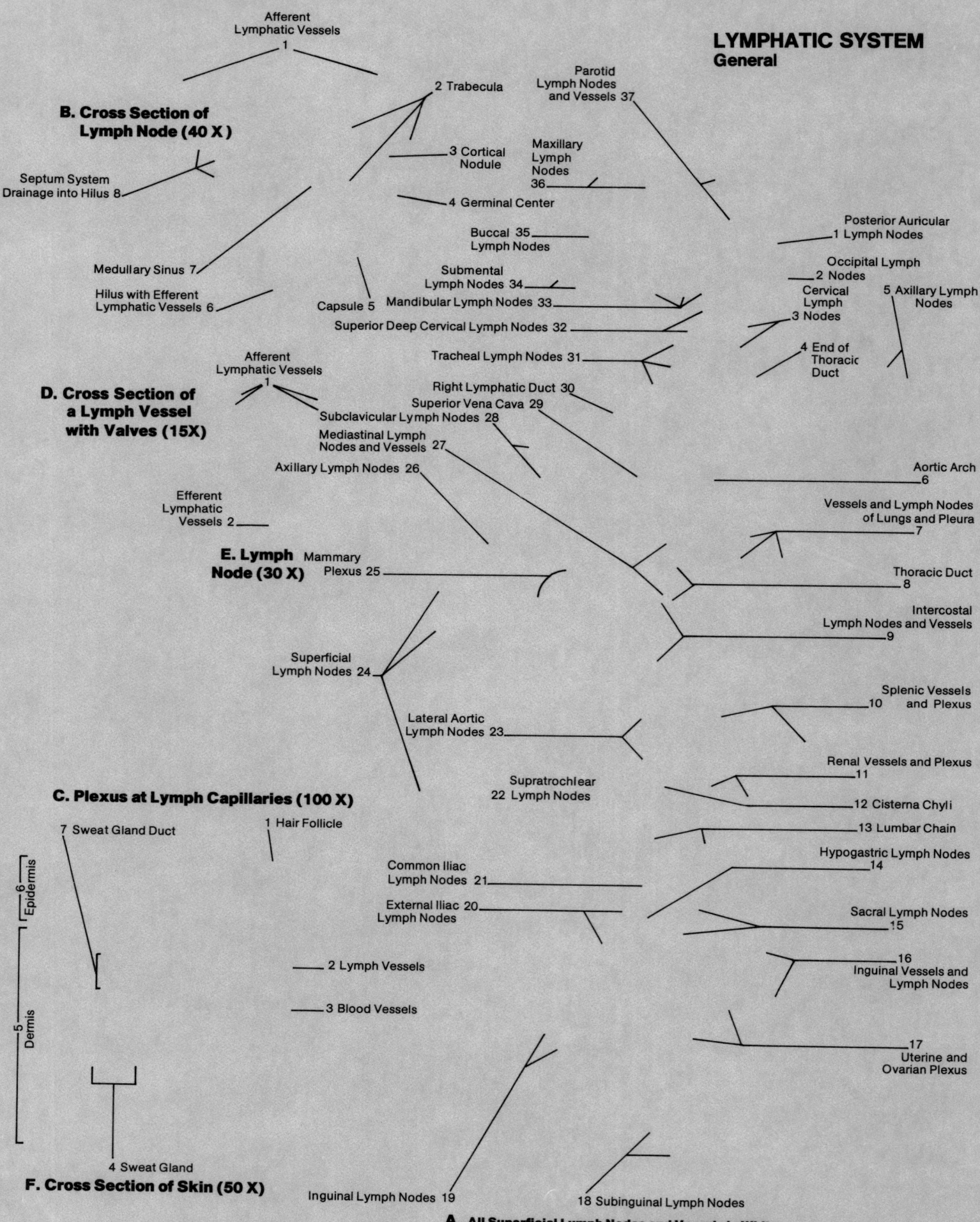

LYMPHATIC SYSTEM
General

Afferent Lymphatic Vessels 1

2 Trabecula

Parotid Lymph Nodes and Vessels 37

B. Cross Section of Lymph Node (40 X)

3 Cortical Nodule

Maxillary Lymph Nodes 36

Septum System Drainage into Hilus 8

4 Germinal Center

Posterior Auricular 1 Lymph Nodes

Buccal 35 Lymph Nodes

Occipital Lymph 2 Nodes

Medullary Sinus 7

Submental Lymph Nodes 34

Cervical Lymph 3 Nodes

5 Axillary Lymph Nodes

Hilus with Efferent Lymphatic Vessels 6

Capsule 5   Mandibular Lymph Nodes 33

Superior Deep Cervical Lymph Nodes 32

4 End of Thoracic Duct

Afferent Lymphatic Vessels 1

Tracheal Lymph Nodes 31

D. Cross Section of a Lymph Vessel with Valves (15X)

Right Lymphatic Duct 30
Superior Vena Cava 29
Subclavicular Lymph Nodes 28

Mediastinal Lymph Nodes and Vessels 27

Aortic Arch 6

Efferent Lymphatic Vessels 2

Axillary Lymph Nodes 26

Vessels and Lymph Nodes of Lungs and Pleura 7

E. Lymph Node (30 X)

Mammary Plexus 25

Thoracic Duct 8

Intercostal Lymph Nodes and Vessels 9

Superficial Lymph Nodes 24

Splenic Vessels 10 and Plexus

Renal Vessels and Plexus 11

Lateral Aortic Lymph Nodes 23

12 Cisterna Chyli

13 Lumbar Chain

C. Plexus at Lymph Capillaries (100 X)

Supratrochlear 22 Lymph Nodes

7 Sweat Gland Duct

1 Hair Follicle

Hypogastric Lymph Nodes 14

6 Epidermis

Common Iliac Lymph Nodes 21

Sacral Lymph Nodes 15

External Iliac 20 Lymph Nodes

16 Inguinal Vessels and Lymph Nodes

2 Lymph Vessels

5 Dermis

3 Blood Vessels

17 Uterine and Ovarian Plexus

4 Sweat Gland

F. Cross Section of Skin (50 X)

Inguinal Lymph Nodes 19

18 Subinguinal Lymph Nodes

A. All Superficial Lymph Nodes and Vessels in White.
All Deep Lymph Nodes and Vessels in Yellow.

NO. NS24 © 1988 AMERICAN MAP CORP

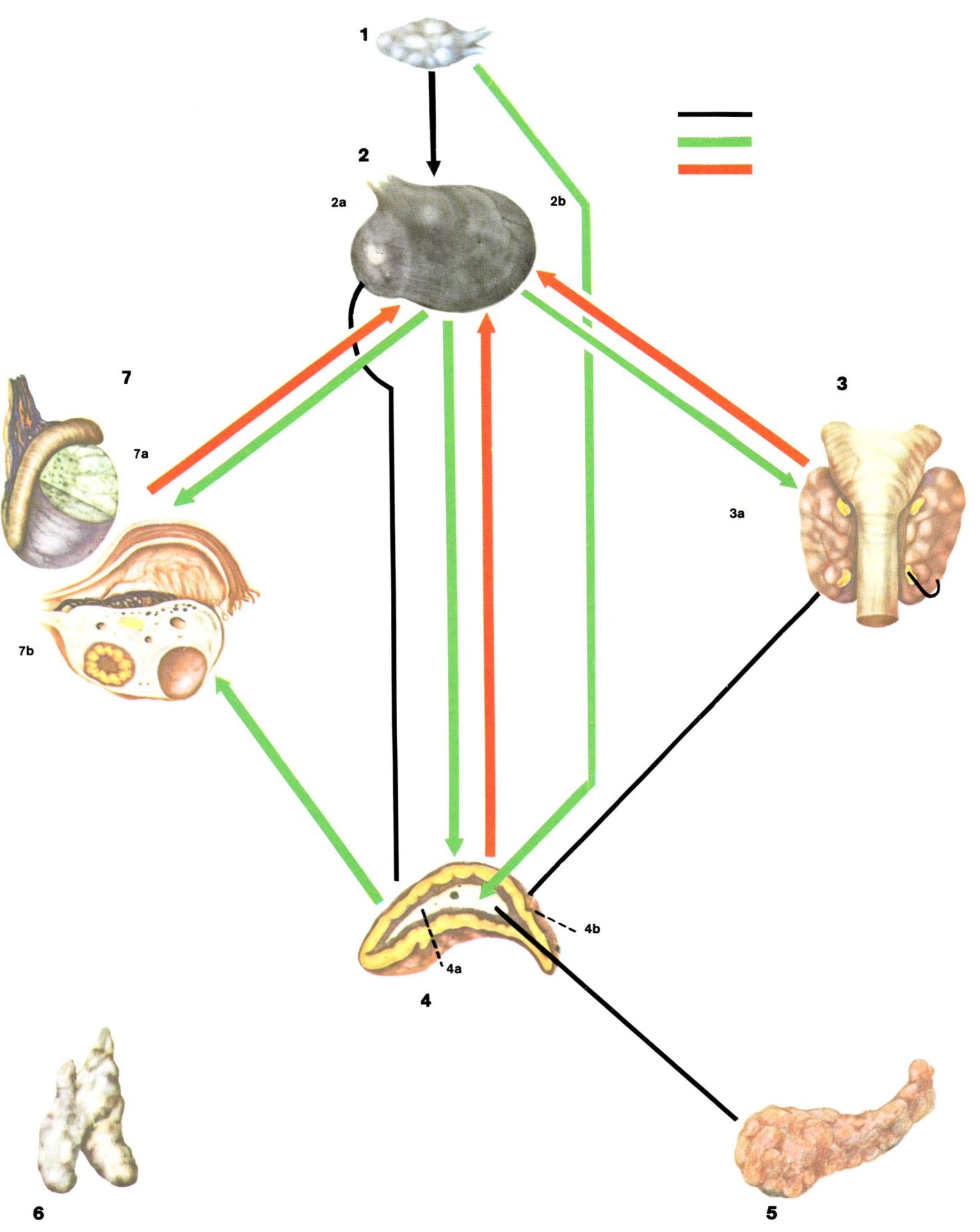

Pineal 1

Antagonism
Stimulation
Feedback Inhibition

Pituitary 2

2b Anterior

Posterior 2a

3. Thyroid

7. Gonads

Testis 7a

Parathyroids 3a

viewed from behind

7b Ovary

4b Cortex

4a Medulla

4. Adrenal

5. Pancreas

6. Thymus

ENDOCRINE GLANDS
Diagram of Endocrine Interrelations

Pineal 1

Antagonism

Stimulation

Feedback Inhibition

Pituitary 2

Posterior 2a

2b Anterior

7. Gonads

3. Thyroid

Testis 7a

Parathyroids 3a

7b Ovary

viewed from behind

4b Cortex

4a Medulla

4. Adrenal

6. Thymus

5. Pancreas

No. NS23 ©1988 AMERICAN MAP CORP.

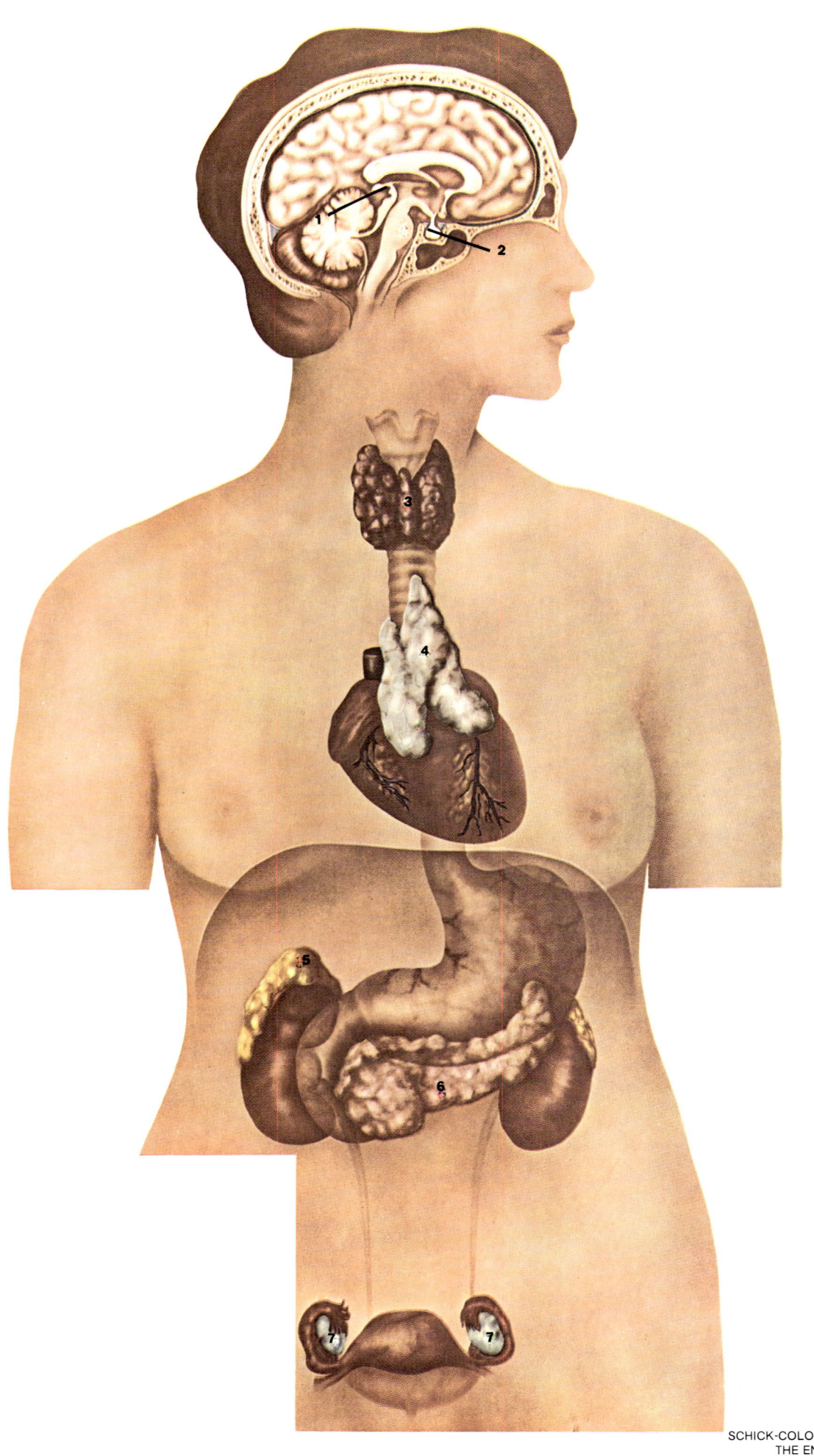

SCHICK-COLORPRINT® ANATOMY CHART
THE ENDOCRINE GLANDS
No. NS22 © 1988 AMERICAN MAP CORP.

3
4
5
6
7

1 Pineal
2 Pituitary

No. NS22 ©1988 AMERICAN MAP CORP.

1

2

3

4

5

6

1 Pineal

2 Pituitary

3 Thyroid

4 Thymus

5 Adrenal   7   7

6 Pancreas

7 Ovaries

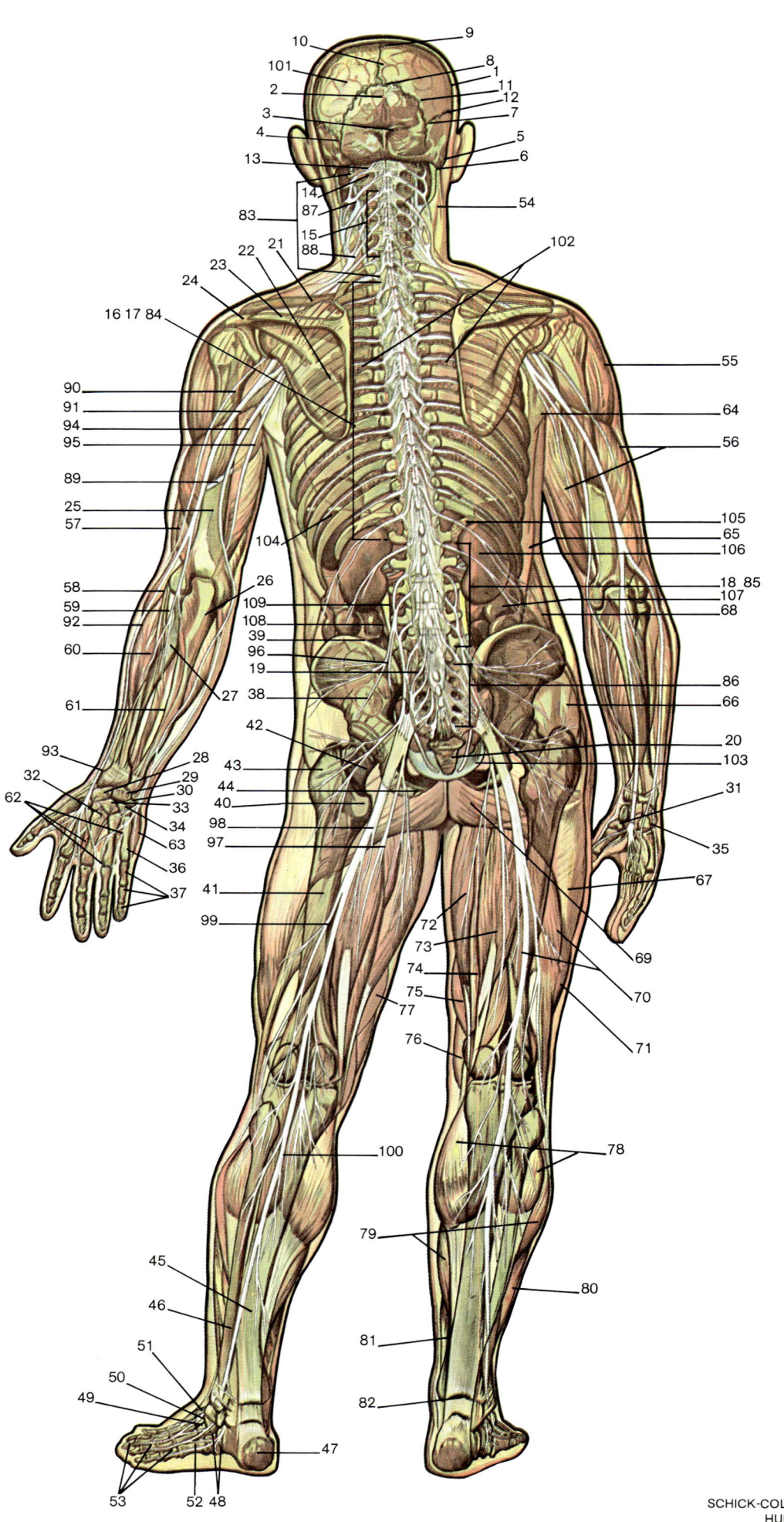

SCHICK-COLORPRINT® ANATOMY CHART
HUMAN BODY-BACK
No. NS21 © 1988 AMERICAN MAP CORP.

79. Soleus
80. Peroneus Longus
81. Flexor Digitorum Longus
82. Calcanean Tendon

Bones

1. Parietal Bone
2. Occipital Bone
3. Inion
4. Temporal Bone
5. Mastoid Process
6. Ascending Ramus of Mandible
7. Asterion
8. Lambda
9. Vertex
10. Sagittal Suture
11. Lambdoidal Suture
12. Squamosal Suture
13. Atlas
14. Axis
15. 5 Cervical Vertebrae
16. 12 Thoracic Vertebrae
17. 12 Ribs
18. 5 Lumbar Vertebrae
19. Sacrum
20. Coccyx
21. Clavical
22. Scapula
23. Spine of Scapula
24. Acromion (Acromial Process)
25. Humerus
26. Ulna
27. Radius
28. Scaphoid Bone
29. Lunate Bone
30. Triquetrum Bone
31. Trapezium
32. Trapezoid
33. Capitate Bone
34. Hamate Bone
35. Pisiform Bone
36. Metacarpals
37. Phalanges
38. Ilium
39. Iliac Crest
40. Ischium
41. Femur
42. Neck of Femur
43. Greater Trochanter
44. Lesser Trochanter
45. Tibia
46. Fibula
47. Calcaneus
48. Cuboid Bone
49. 3rd Cuneiform Bone
50. 2nd Cuneiform Bone
51. 1st Cuneiform Bone
52. Metatarsals
53. Phalanges

Nerves

83. 7 Cervical Nerves
84. 12 Thoracic Nerves
85. 5 Lumbar Nerves
86. 5 Sacral Nerves
87. Cervical Plexus
88. Brachial Plexus
89. Radial Nerve
90. Median Nerve
91. Posterior Brachial Cutaneous
92. Deep Radial
93. Deep Branch of Radial
94. Ulnar
95. Medial Brachial Cutaneous
96. Lateral Femoral Cutaneous
97. Posterior Femoral Cutaneous
98. Sciatic
99. Common Peroneal
100. Tibial

Viscera

101. Cerebrum
102. Lungs, Right and Left
103. Bladder
104. Spleen
105. Suprarenal (Adrenal) Glands, Right and Left
106. Kidneys, Right and Left
107. Ascending Colon
108. Descending Colon
109. Ureters, Right and Left

Bones

1. Parietal Bone
2. Occipital Bone
3. Inion
4. Temporal Bone
5. Mastoid Process
6. Ascending Ramus of Mandible
7. Asterion
8. Lambda
9. Vertex
10. Sagittal Suture
11. Lambdoidal Suture
12. Squamosal Suture
13. Atlas
14. Axis
15. 5 Cervical Vertebrae
16. 12 Thoracic Vertebrae
17. 12 Ribs
18. 5 Lumbar Vertebrae
19. Sacrum
20. Coccyx
21. Clavical
22. Scapula
23. Spine of Scapula
24. Acromion (Acromial Process)
25. Humerus
26. Ulna
27. Radius
28. Scaphoid Bone
29. Lunate Bone
30. Triquetrum Bone
31. Trapezium
32. Trapezoid
33. Capitate Bone
34. Hamate Bone
35. Pisiform Bone
36. Metacarpals
37. Phalanges
38. Ilium
39. Iliac Crest
40. Ischium
41. Femur
42. Neck of Femur
43. Greater Trochanter
44. Lesser Trochanter
45. Tibia
46. Fibula
47. Calcaneus
48. Cuboid Bone
49. 3rd Cuneiform Bone
50. 2nd Cuneiform Bone
51. 1st Cuneiform Bone
52. Metatarsals
53. Phalanges

Muscles

54. Sternocleidomastoid
55. Deltoid
56. Triceps
57. Brachialis
58. Brachioradialis
59. Extensor Digitorum Longus
60. Extensor Carpi Radialis
61. Extensor Carpi Ulnaris
62. Dorsal Interossei
63. Opponens Digiti Minimi
64. Teres Major
65. Latissimus Dorsi
66. Gluteus Medius
67. Tensor Fasciae Latae
68. External Oblique
69. Gluteus Maximus
70. Biceps Femoris
71. Vastus Lateralis
72. Adductor Magnus
73. Semimembranosis
74. Semitendonosis
75. Gracilis
76. Sartorius
77. Vastus Medialis
78. Gastrocnemius
79. Soleus
80. Peroneus Longus
81. Flexor Digitorum Longus
82. Calcanean Tendon

Nerves

83. 7 Cervical Nerves
84. 12 Thoracic Nerves
85. 5 Lumbar Nerves
86. 5 Sacral Nerves
87. Cervical Plexus
88. Brachial Plexus
89. Radial Nerve
90. Median Nerve
91. Posterior Brachial Cutaneous
92. Deep Radial
93. Deep Branch of Radial
94. Ulnar
95. Medial Brachial Cutaneous
96. Lateral Femoral Cutaneous
97. Posterior Femoral Cutaneous
98. Sciatic
99. Common Peroneal
100. Tibial

Viscera

101. Cerebrum
102. Lungs, Right and Left
103. Bladder
104. Spleen
105. Suprarenal (Adrenal) Glands, Right and Left
106. Kidneys, Right and Left
107. Ascending Colon
108. Descending Colon
109. Ureters, Right and Left

No. NS21 © 1988 AMERICAN MAP CORP.

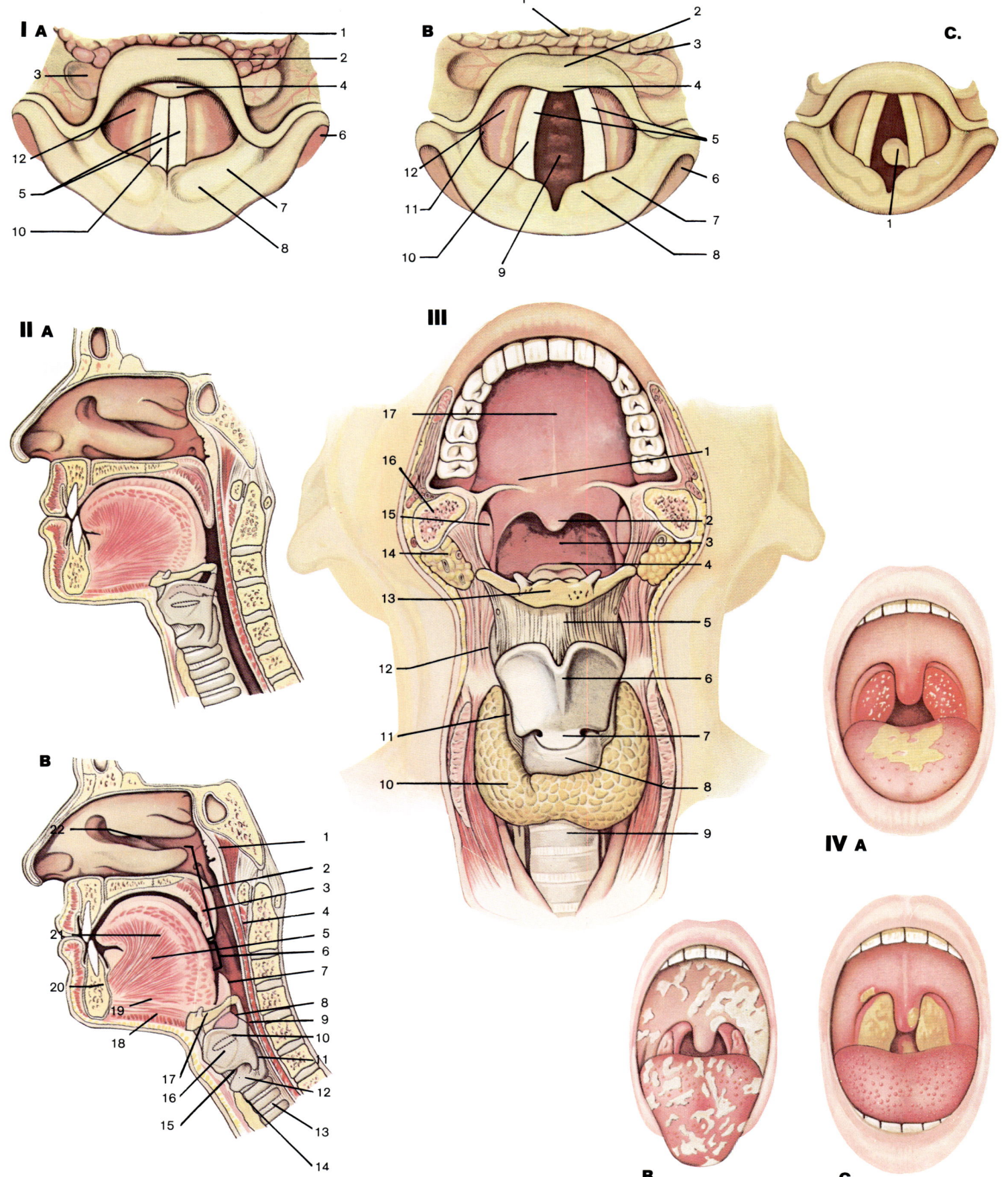

SCHICK-COLORPRINT® ANATOMY CHART
THE THROAT
No. NS20 © 1988 AMERICAN MAP CORP.

4
3
6
12
5
10
7
Tonsil 15
2 Uvula
Submaxillary Gland 14
3 Pharynx
4 Epiglottis
Hyoid Bone 13
5 Thyrohyoid Membrane
Superior Horn of Thyroid Cartilage 12
6 Thyroid Cartilage
Inferior Horn of Thyroid Cartilage 11
7 Cricothyroid Membrane
Thyroid Gland 10
8 Cricoid Cartilage
9 Trachea

IV A. Tonsillitis

B. Position of the Larynx and Pharynx in Breathing

1 Pharyngeal Tonsil
2 Nasopharynx
3 Soft Palate
4 Pharynx
5 Tongue
6 Oropharynx
7 Epiglottis
8 Entrance to Larynx
9 Superior Horn of Thyroid Cartilage
10 Position of Vocal Cords indicated by dashes
11 Inferior Horn of Thyroid Cartilage
12 Cricoid Cartilage
13 Trachea
14 Thyroid Gland

Nasal Cavity 22
Tongue 21
Mandible 20
Geniohyoid Muscle 19
Mylohyoid Muscle 18
Hyoid Bone 17
Thyroid Cartilage 16
Cricothyroid Membrane 15

B. Thrush (Candida Albicans)

C. Diphtheria

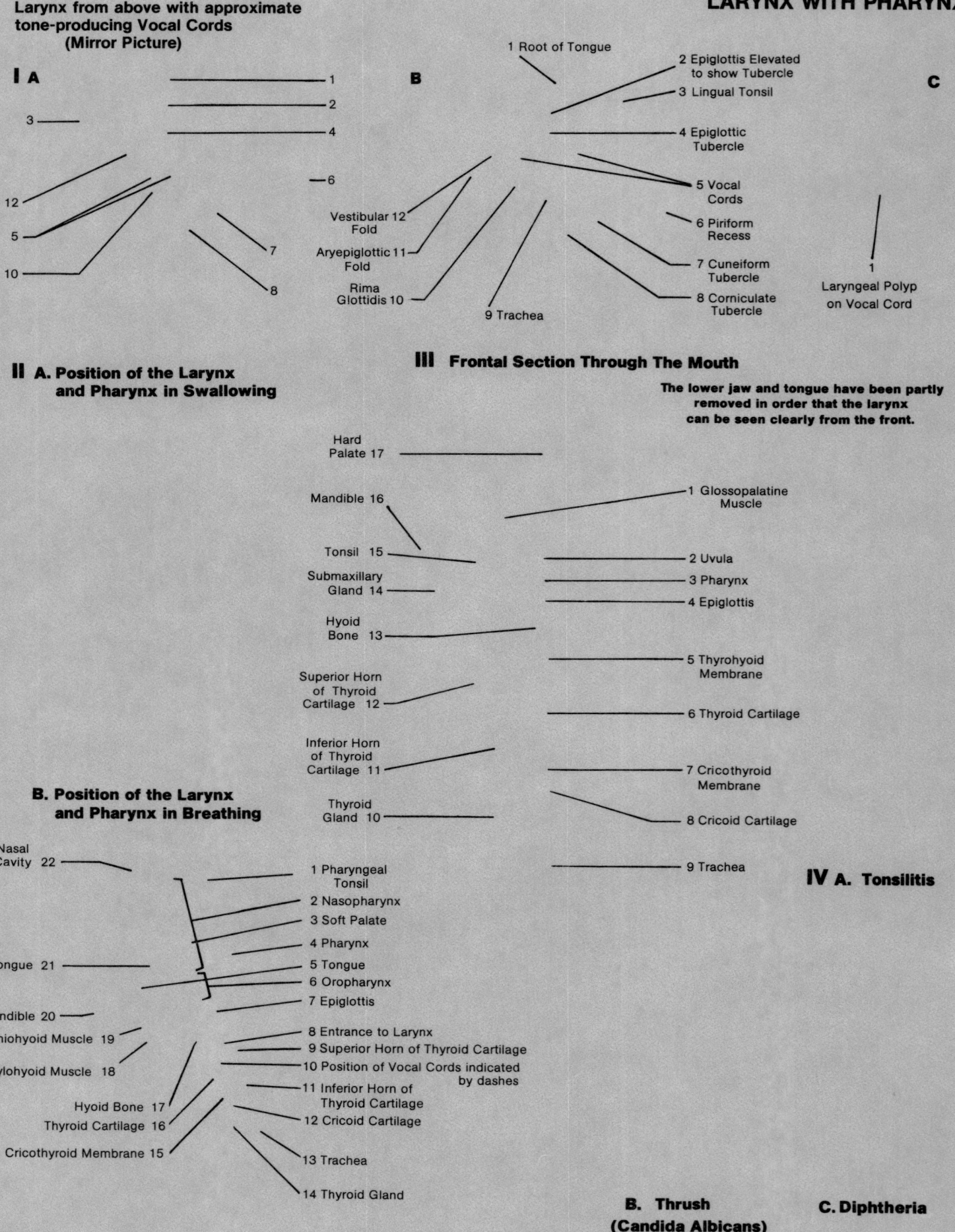

LARYNX WITH PHARYNX

Larynx from above with approximate tone-producing Vocal Cords (Mirror Picture)

I A

1
2
4
3
6
12
5
10
7
8

B

1 Root of Tongue
2 Epiglottis Elevated to show Tubercle
3 Lingual Tonsil
4 Epiglottic Tubercle
5 Vocal Cords
6 Piriform Recess
7 Cuneiform Tubercle
8 Corniculate Tubercle

Vestibular 12 Fold
Aryepiglottic 11 Fold
Rima Glottidis 10
9 Trachea

C

1
Laryngeal Polyp on Vocal Cord

II A. Position of the Larynx and Pharynx in Swallowing

III Frontal Section Through The Mouth

The lower jaw and tongue have been partly removed in order that the larynx can be seen clearly from the front.

Hard Palate 17
Mandible 16
Tonsil 15
Submaxillary Gland 14
Hyoid Bone 13
Superior Horn of Thyroid Cartilage 12
Inferior Horn of Thyroid Cartilage 11
Thyroid Gland 10

1 Glossopalatine Muscle
2 Uvula
3 Pharynx
4 Epiglottis
5 Thyrohyoid Membrane
6 Thyroid Cartilage
7 Cricothyroid Membrane
8 Cricoid Cartilage
9 Trachea

B. Position of the Larynx and Pharynx in Breathing

Nasal Cavity 22
Tongue 21
Mandible 20
Geniohyoid Muscle 19
Mylohyoid Muscle 18
Hyoid Bone 17
Thyroid Cartilage 16
Cricothyroid Membrane 15

1 Pharyngeal Tonsil
2 Nasopharynx
3 Soft Palate
4 Pharynx
5 Tongue
6 Oropharynx
7 Epiglottis
8 Entrance to Larynx
9 Superior Horn of Thyroid Cartilage
10 Position of Vocal Cords indicated by dashes
11 Inferior Horn of Thyroid Cartilage
12 Cricoid Cartilage
13 Trachea
14 Thyroid Gland

IV A. Tonsilitis

B. Thrush (Candida Albicans)

C. Diphtheria

No. NS20 © 1988 AMERICAN MAP CORP.

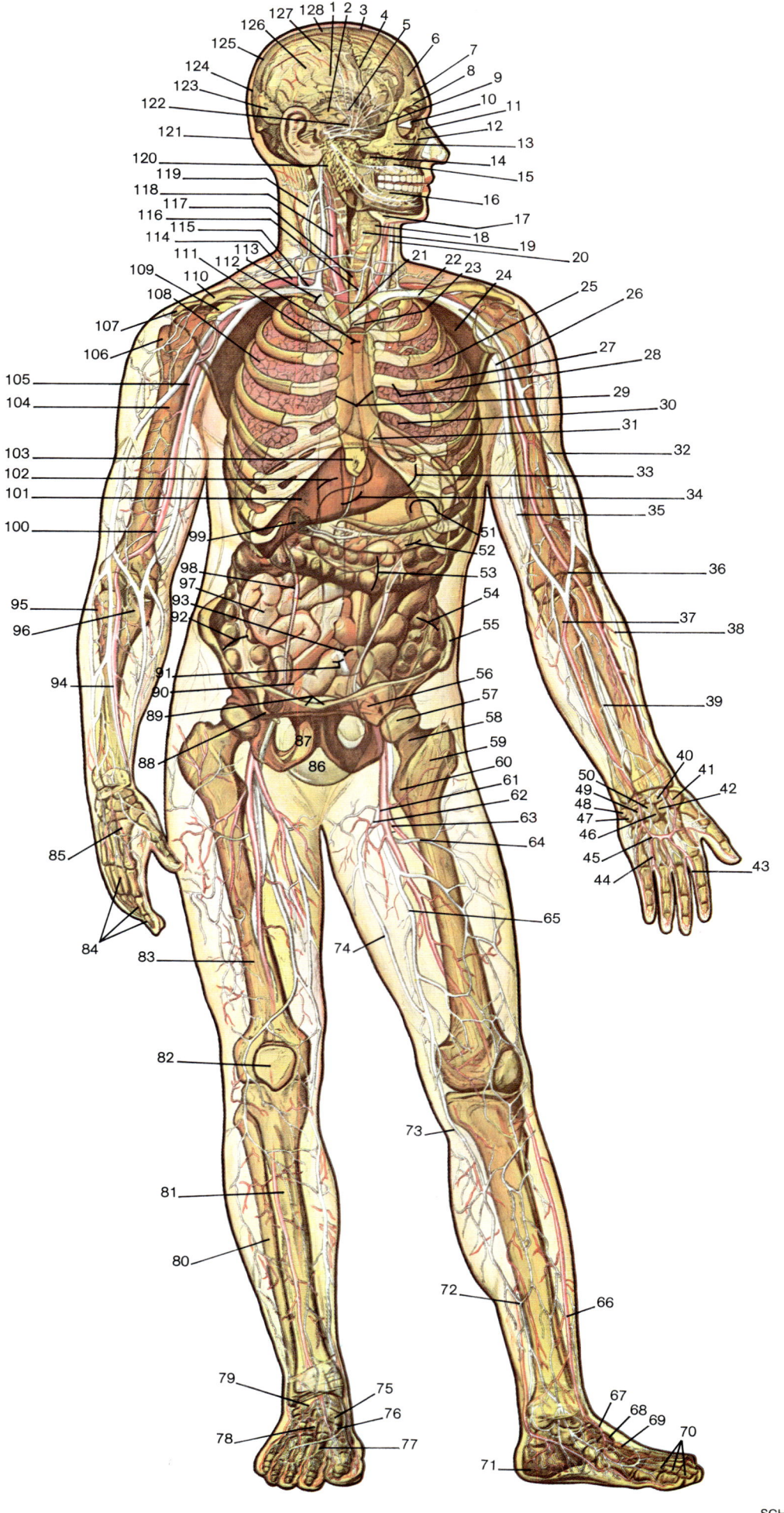

27. Ribs
28. Costal Cartilages
29. Body of Sternum
30. Heart
31. Esophagus
32. Cephalic Vein
33. Stomach
34. Abdominal Aorta
35. Basilic Vein
36. Median Cubital Vein
37. Ulnar Artery
38. Accessory Cephalic Vein
39. Median Antebrachial Vein
40. Scaphoid Bone
41. Trapezium
42. Trapezoid
43. Proper Palmar Digital Artery
44. Common Palmar Digital Artery
45. Superficial Palmar Arch
46. Capitate Bone
47. Triquetrum Bone
48. Pisiform Bone
49. Hamate Bone
50. Lunate Bone
51. Kidney
52. Pancreas
53. Transverse Colon
54. Descending Colon
55. Ilium
56. Pubis
57. Head of Femur
58. Neck of Femur
59. Greater Trochanter
60. Lesser Trochanter
61. Femoral Artery
62. Femoral Vein
63. Lateral Femoral Circumflex Artery
64. Lateral Femoral Circumflex Vein

65. Femoral Vein
66. Anterior Tibial Artery
67. Dorsalis Pedis Artery
68. Third Cuneiform Bone
69. Second Cuneiform Bone
70. Phalanges
71. Calcaneus
72. Posterior Tibial Artery
73. Great Saphenous Vein
74. Great Saphenous Vein
75. Navicular Bone
76. First Cuneiform Bone
77. Metatarsals
78. Cuboid Bone
79. Talus
80. Fibula
81. Tibia
82. Patella
83. Femur
84. Phalanges
85. Metacarpals
86. Bladder
87. Ischium
88. Vermiform Appendix
89. Pubic Symphysis
90. Internal Iliac Artery
91. Common Iliac Vein
92. Ascending Colon
93. Common Iliac Artery
94. Radial Artery
95. Radius
96. Ulna
97. Small Intestine
98. Ureter
99. Gall Bladder
100. Brachial Artery
101. Liver
102. Inferior Vena Cava
103. Xiphoid Process
104. Humerus
105. Axillary Artery
106. Intertubercular Sulcus
107. Acromion (Acromial Process)
108. Right Lung
109. Coracoid Process of Scapula
110. Clavicle
111. Superior Vena Cava
112. Arch of the Aorta
113. Right Brachiocephalic Vein
114. Subclavian Artery
115. Brachiocephalic Trunk
116. Inferior Thyroid Vein
117. Superior Thyroid Artery
118. Carotid Artery
119. External Jugular Vein
120. Parotid Gland
121. Inion
122. Gasserion Ganglion and Branches of 5th Cranial Nerve
123. Occipital Bone
124. Lambda
125. Obelion
126. Parietal Bone
127. Superior Temporal Line
128. Vertex

Human Body with all Organs and Circulation of Blood

HUMAN BODY - Front

1. Inferior Temporal Line
2. Temporal Bone
3. Bregma
4. Stephanion
5. Pterion
6. Frontal Bone
7. Ophryon
8. Glabella
9. Sphenoid Bone
10. Nasion
11. Nasal Bone
12. Lacrimal Bone
13. Zygomatic Bone
14. Coronoid Process of Mandible
15. Maxilla
16. Mandible
17. Hyoid Bone
18. Thyroid Cartilage
19. Trachea
20. Internal Jugular Vein
21. Left Bracheocephalic Vein
22. Subclavian Vein
23. Manubrium of Sternum
24. Scapula
25. Left Lung
26. Axillary Vein
27. Ribs
28. Costal Cartilages
29. Body of Sternum
30. Heart
31. Esophagus
32. Cephalic Vein
33. Stomach
34. Abdominal Aorta
35. Basilic Vein
36. Median Cubital Vein
37. Ulnar Artery
38. Accessory Cephalic Vein
39. Median Antebrachial Vein
40. Scaphoid Bone
41. Trapezium
42. Trapezoid
43. Proper Palmar Digital Artery
44. Common Palmar Digital Artery
45. Superficial Palmar Arch
46. Capitate Bone
47. Triquetrum Bone
48. Pisiform Bone
49. Hamate Bone
50. Lunate Bone
51. Kidney
52. Pancreas
53. Transverse Colon
54. Descending Colon
55. Ilium
56. Pubis
57. Head of Femur
58. Neck of Femur
59. Greater Trochanter
60. Lesser Trochanter
61. Femoral Artery
62. Femoral Vein
63. Lateral Femoral Circumflex Artery
64. Lateral Femoral Circumflex Vein
65. Femoral Vein
66. Anterior Tibial Artery
67. Dorsalis Pedis Artery
68. Third Cuneiform Bone
69. Second Cuneiform Bone
70. Phalanges
71. Calcaneus
72. Posterior Tibial Artery
73. Great Saphenous Vein
74. Great Saphenous Vein
75. Navicular Bone
76. First Cuneiform Bone
77. Metatarsals
78. Cuboid Bone
79. Talus
80. Fibula
81. Tibia
82. Patella
83. Femur
84. Phalanges
85. Metacarpals
86. Bladder
87. Ischium
88. Vermiform Appendix
89. Pubic Symphysis
90. Internal Iliac Artery
91. Common Iliac Vein
92. Ascending Colon
93. Common Iliac Artery
94. Radial Artery
95. Radius
96. Ulna
97. Small Intestine
98. Ureter
99. Gall Bladder
100. Brachial Artery
101. Liver
102. Inferior Vena Cava
103. Xiphoid Process
104. Humerus
105. Axillary Artery
106. Intertubercular Sulcus
107. Acromion (Acromial Process)
108. Right Lung
109. Coracoid Process of Scapula
110. Clavicle
111. Superior Vena Cava
112. Arch of the Aorta
113. Right Bracheocephalic Vein
114. Subclavian Artery
115. Bracheocephalic Trunk
116. Inferior Thyroid Vein
117. Superior Thyroid Artery
118. Carotid Artery
119. External Jugular Vein
120. Parotid Gland
121. Inion
122. Gasserion Ganglion and Branches of 5th Cranial Nerve
123. Occipital Bone
124. Lambda
125. Obelion
126. Parietal Bone
127. Superior Temporal Line
128. Vertex

Human Body with all Organs and Circulation of Blood

No. NS19 © 1988 AMERICAN MAP CORP.

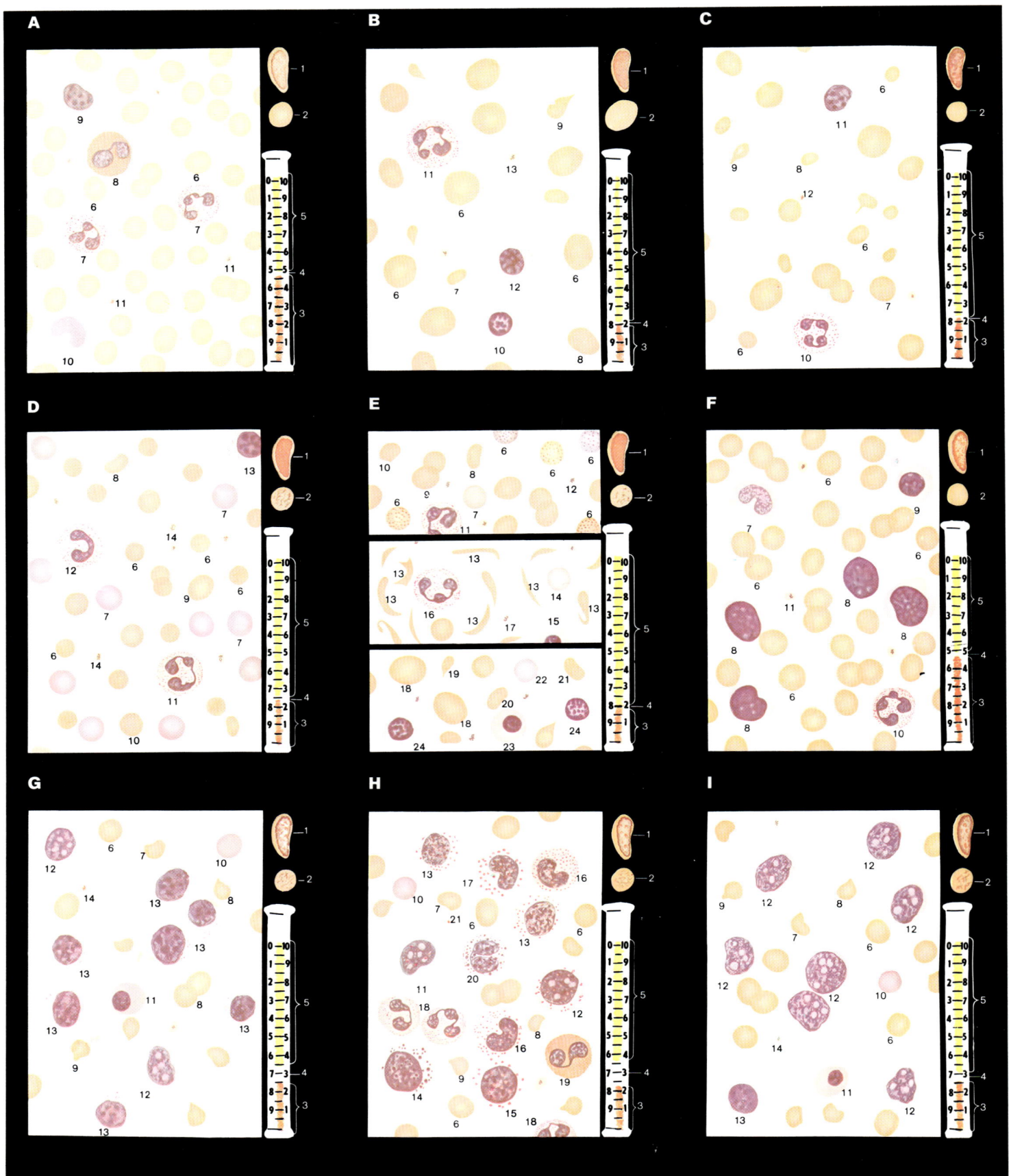

SCHICK-COLOR PRINT® ANATOMY CHART
DISEASES OF THE BLOOD CELLS
No. NS18 ©1988 AMERICAN MAP CORP.

C. Microcytic Hypochromic Anemia (Iron Deficiency)

1. Bone marrow: hyperplasia
2. Red blood cell: microcytic, hypochromic, decreased regeneration
3. Volume of packed red blood cells: moderately decreased
4. Volume of packed white blood cells: normal
5. Plasma volume: moderately increased
6. Microcyte
7, 8, 9. Variation in size and shape of the red blood cells
10. Neutrophil
11. Lymphocyte
12. Platelet

D. Congenital Hemolytic Anemia

1. Bone marrow: hyperplasia
2. Red blood cell: normocytic
3. Volume of packed red blood cells: moderately decreased
4. Volume of packed white blood cells: normal
5. Plasma volume: moderately increased, icteric
6. Spherical microcyte
7. Polychromatophilia
8, 9, 10. Variation in size and shape of the red blood cells
11. Neutrophil
12. Unsegmented Neutrophil
13. Lymphocyte
14. Platelet

E. Special Forms of Anemia

1. Bone marrow: hyperplasia
2. Red blood cell: normocytic, increased regeneration
3. Volume of packed red blood cells: decreased
4. Volume of packed white blood cells: normal
5. Plasma volume: moderately increased

Upper Third: Lead Poisoning
6. Stippled red blood cells
7. Polychromatophilia
8, 9, 10. Variation in size and shape of the red blood cells
11. Neutrophil
12. Platelet

Middle Third: Sickle Cell Anemia
13. Sickle shaped red blood cells
14. Polychromatophilia
15. Normoblast
16. Neutrophil
17. Platelet

Lower Third: Erythroblastosis Fetalis
18. Macrocyte
19, 20, 21. Variation in size and shape of the red blood cells
22. Polychromatophilia
23. Normoblast
24. Erythroblast

F. Infectious Mononucleosis

1. Bone marrow: normal
2. Red blood cell: normal
3. Volume of packed red blood cells: normal
4. Volume of packed white blood cells: normal
5. Plasma volume: normal
6. Normal red blood cell
7. Monocyte
8. Atypical lymphocyte
9. Lymphocyte
10. Neutrophil
11. Platelet

G. Chronic Lymphocytic Leukemia

1. Bone marrow: marrow infiltration, decrease in normal hematopoietic stem cells
2. Red blood cell: normal, increased regeneration
3. Volume of packed red blood cells: decreased
4. Volume of packed white blood cells: markedly increased
5. Plasma volume: slightly increased
6. Normal red blood cell
7, 8, 9. Variation in size and shape of the red blood cells
10. Polychromatophilia
11. Normoblast
12. Lymphoblast
13. Lymphocyte
14. Platelet

H. Chronic Myelocytic Leukemia

1. Bone marrow: marrow infiltration hyperplasia
2. Red blood cell: normal, increased regeneration
3. Volume of packed red blood cells: decreased
4. Volume of packed white blood cells: markedly increased
5. Plasma volume: increased
6. Normal red blood cell
7, 8, 9. Variation in size and shape of the red blood cells
10. Polychromatophilia
11. Myeloblast
12, 13. Neutrophilic myelocytes
14. Basophilic myelocyte
15. Eosinophilic myelocyte
16. Neutrophilic metamyelocyte
17. Eosinophilic metamyelocyte
18. Neutrophil
19. Eosinophil
20. Basophil
21. Platelet

I. Acute Leukemia

1. Bone marrow: marrow infiltration
2. Red blood cell: normal, increased regeneration
3. Volume of packed red blood cells: decreased
4. Volume of packed white blood cells: increased
5. Plasma volume: increased
6. Normal red blood cell
7, 8, 9. Variation in size and shape of red blood cells
10. Polychromatophilia
11. Normoblast
12. Stem cell
13. Lymphocyte
14. Platelet

DISEASES OF THE BLOOD CELLS

A. Normal Blood

1. Bone marrow: normal
2. Red blood cell: normal, normal regeneration
3. Volume of packed red blood cells: 47%
4. Volume of packed white blood cells: 1%
5. Plasma volume: 52%
6. Normal red blood cell
7. Neutrophil
8. Eosinophil
9. Lymphocyte
10. Monocyte
11. Platelet

B. Megaloblastic Anemia (Pernicious Anemia)

1. Bone marrow: hyperplasia
2. Red blood cell: macrocytic, hyperchromic, decreased regeneration
3. Volume of packed red blood cells: markedly decreased
4. Volume of packed white blood cells slightly decreased
5. Plasma volume: marked increase slighly icteric
6. Macrocyte
7, 8, 9. Variation in size and shape of the red blood cells
10. Megaloblast
11. Neutrophil
12. Lymphocyte
13. Platelet

C. Microcytic Hypochromic Anemia (Iron Deficiency)

1. Bone marrow: hyperplasia
2. Red blood cell: microcytic, hypochromic, decreased regeneration
3. Volume of packed red blood cells: moderately decreased
4. Volume of packed white bood cells: normal
5. Plasma volume: moderately increased
6. Microcyte
7, 8, 9. Variation in size and shape of the red blood cells
10. Neutrophil
11. Lymphocyte
12. Platelet

D. Congenital Hemolytic Anemia

1. Bone marrow: hyperplasia
2. Red blood cell: normocytic
3. Volume of packed red blood cells: moderately decreased
4. Volume of packed white blood cells: normal
5. Plasma volume: moderately increased, icteric
6. Spherical microcyte
7. Polychromatophilia
8, 9, 10. Variation in size and shape of the red blood cells
11. Neutrophil
12. Unsegmented Neutrophil
13. Lymphocyte
14. Platelet

E. Special Forms of Anemia

1. Bone marrow: hyperplasia
2. Red blood cell: normocytic, increased regeneration
3. Volume of packed red blood cells: decreased
4. Volume of packed white blood cells: normal
5. Plasma volume: moderately increased
Upper Third: Lead Poisoning
6. "Stippled" red blood cells
7. Polychromatophilia
8, 9, 10. Variation in size and shape of the red blood cells
11. Neutrophil
12. Platelet
Middle Third: Sickle Cell Anemia
13. Sickle shaped red blood cells
14. Polychromatophilia
15. Normoblast
16. Neutrophil
17. Platelet
Lower Third: Erythroblastosis Fetalis
18. Macrocyte
19, 20, 21. Variation in size and shape of the red blood cells
22. Polychromatophilia
23. Normoblast
24. Erythroblast

F. Infectious Mononucleosis

1. Bone marrow: normal
2. Red blood cell: normal
3. Volume of packed red blood cells: normal
4. Volume of packed white blood cells: normal
5. Plasma volume: normal
6. Normal red blood cell
7. Monocyte
8. Atypical lymphocyte
9. Lymphocyte
10. Neutrophil
11. Platelet

G. Chronic Lymphocytic Leukemia

1. Bone marrow: marrow infiltration, decrease in normal hematopoietic stem cells
2. Red blood cell: normal, increased regeneration
3. Volume of packed red blood cells: decreased
4. Volume of packed white blood cell: markedly increased
5. Plasma volume: slightly increased
6. Normal red bood cell
7, 8, 9. Variation in size and shape of the red blood cells
10. Polychromatophilia
11. Normoblast
12. Lymphoblast
13. Lymphocyte
14. Platelet

H. Chronic Myelocytic Leukemia

1. Bone marrow: marrow infiltration hyperplasia
2. Red blood cell: normal, increased regeneration
3. Volume of packed red blood cells: decreased
4. Volume of packed white blood cells: markedly increased
5. Plasma volume: increased
6. Normal red blood cell
7, 8, 9. Variation in size and shape of the red blood cells
10. Polychromatophilia
11. Myeloblast
12, 13. Neutrophilic myelocytes
14. Basophilic myelocyte
15. Eosinophilic myelocyte
16. Neutrophilic metamyelocyte
17. Eosinophilic metamyelocyte
18. Neutrophil
19. Eosinophil
20. Basophil
21. Platelet

I. Acute Leukemia

1. Bone marrow: marrow infiltration
2. Red blood cell: normal, increased regeneration
3. Volume of packed red blood cells: decreased
4. Volume of packed white blood cells: increased
5. Plasma volume: increased
6. Normal red blood cell
7, 8, 9. Variation in size and shape of red blood cells
10. Polychromatophilia
11. Normoblast
12. Stem cell
13. Lymphocyte
14. Platelet

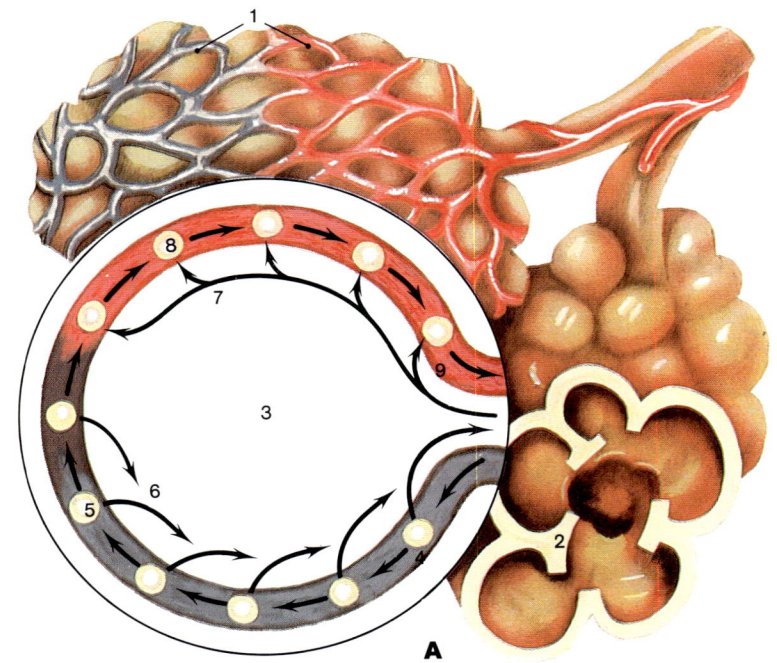

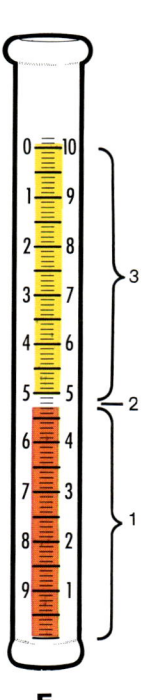

A

E

B

C

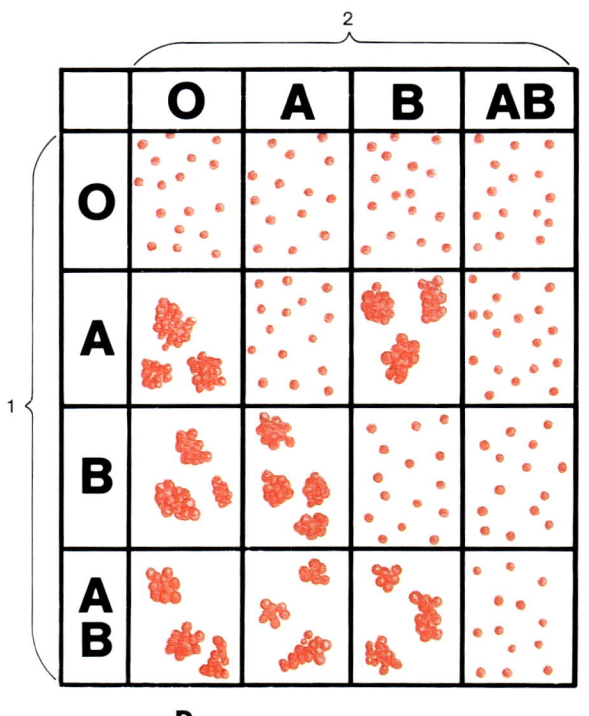

D

SCHICK-COLORPRINT® ANATOMY CHART
BLOOD CELLS
No. NS17 © 1988 AMERICAN MAP CORP.

BLOOD CELLS

A. Function of the Red Blood Cells

1 Alveolus and Capillary Network

2 Cross-Section of Alveolus, showing Alveolar Sac

3 Schematic Representation, showing Exchange of Oxygen and Carbon Dioxide

4 Deoxygenated Blood moving through a Capillary

5 Carbon Dioxide is released from Red Blood Cells

6 Carbon Dioxide passes into the Alveoli and is expired

7 Oxygen of the inspired air passes through the Alveolar Wall

8 Oxygen enters Red Blood Cells and is carried to all Organs

9 Alveolar Capillary

E. Normal Blood Cell Volume
(Hematocrit Tube)

3 Plasma (Serum): 52%

2 Leukocytes (White Blood Cells) & Platelets (Thrombocytes): 1%

1 Erythrocytes (Red Blood Cells): 47%

Normal Blood Counts

A. Function of the Red Blood Cells

1 Alveolus and Capillary Network

2 Cross-Section of Alveolus, showing Alveolar Sac

3 Schematic Representation, showing Exchange of Oxygen and Carbon Dioxide

4 Deoxygenated Blood moving through a Capillary

5 Carbon Dioxide is released from Red Blood Cells

6 Carbon Dioxide passes into the Alveoli and is expired

7 Oxygen of the inspired air passes through the Alveolar Wall

8 Oxygen enters Red Blood Cells and is carried to all Organs

9 Alveolar Capillary

E. Normal Blood Cell Volume
(Hematocrit Tube)

3 Plasma (Serum): 52%

2 Leukocytes (White Blood Cells) & Platelets (Thrombocytes): 1%

1 Erythrocytes (Red Blood Cells): 47%

Normal Blood Counts

Erythrocytes (Red Cells): 4,500,000 - 5,000,000/cu.mm.blood
Thrombocytes (Platelets): 200,000 - 400,000/cu.mm.blood
Leukocytes (White Cells): 5,000-10,000/cu.mm.blood

Neutrophils: 55 - 60%
Lymphocytes: 30 - 35%
Monocytes: 4 - 8%
Eosinophils: 2 - 5%
Basophils: 0 - 1.5%

Hemoglobin

Men: 14.5 - 16.5gm/100ml blood
Women: 12.7 - 14.7gm/100ml blood

1 Foreign Body

2 Bacteria on Foreign Body

3 White Cells moving toward Bacteria

4 Arterial Capillary

5 Venous Capillary

2 Leukocyte

1 Endothelial Cell

3 Migration of Leukocyte through the Capillary Wall

2
Recipient Blood Group

Donor Blood Group

1

B. Function of the White Blood Cells
(Schematic Cross-Section of Skin)

C. Microscopic Section of Inflamed Capillary Blood Vessel

D. Compatibility of Blood Groups
(Large Red Clumps mean Incompatibility)

No. NS17 ©1988 AMERICAN MAP CORP

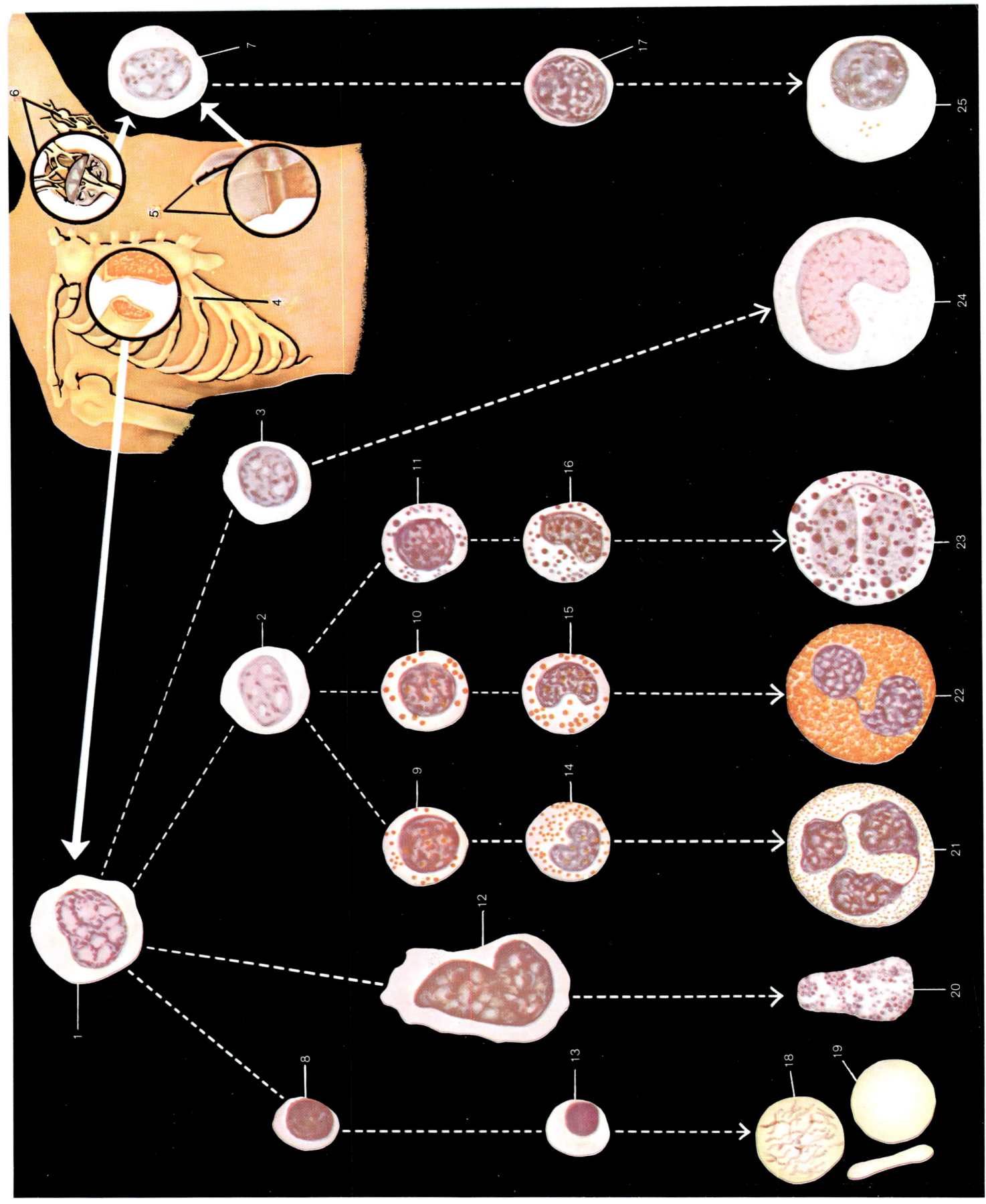

SCHICK-COLORPRINT® ANATOMY CHART
DEVELOPMENT OF BLOOD CELLS
No. NS16 ©1988 AMERICAN MAP CORP

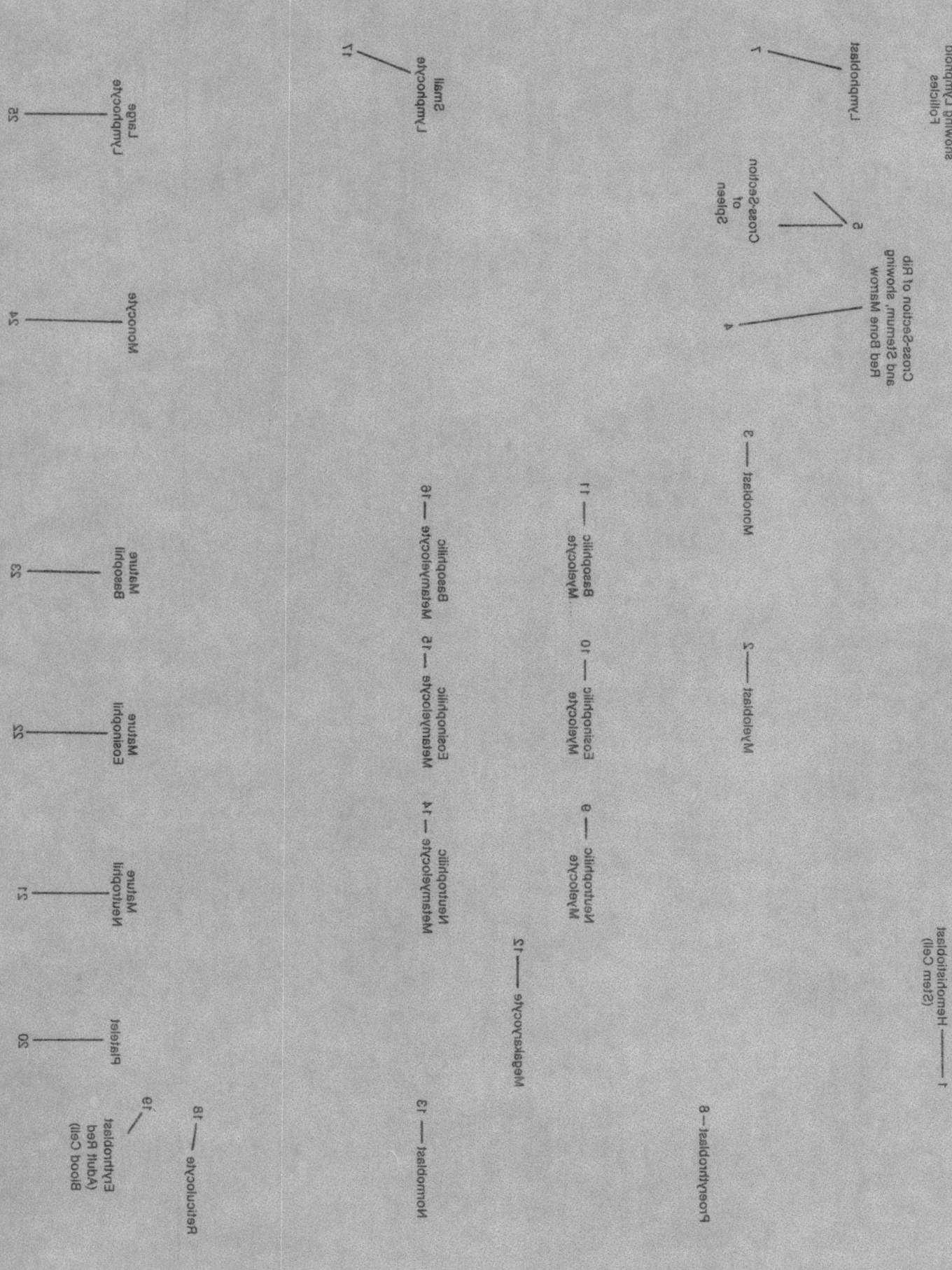

Lymphoblast

7

Cross-Section
of
Spleen

8 —— Lymphoblast

Large
Lymphocyte
25

Small
Lymphocyte
17

Monocyte
24

Monoblast ——— 3

Myeloblast ——— 2

Basophilic
Myelocyte ——— 11

Basophilic
Metamyelocyte ——— 16

Mature
Basophil
Myelocyte ——— 10

Eosinophilic
Metamyelocyte ——— 15

Mature
Eosinophil
22

Neutrophilic
Myelocyte ——— 9

Neutrophilic
Metamyelocyte ——— 14

Mature
Neutrophil
21

Megakaryocyte ——— 12

Platelet
20

Normoblast ——— 13

Reticulocyte ——— 18

Erythroblast
(Adult Red
Blood Cell)
19

DEVELOPMENT OF THE BLOOD CELLS

6

Cross-Section of
Axillary Lymph Node
showing Lymphoid
Follicles

Lymphoblast

7

Cross-Section of Rib
and Sternum, showing
Red Bone Marrow

5

Cross-Section
of
Spleen

4

Small
Lymphocyte

17

Large
Lymphocyte

25

Monocyte

24

1 —— Hemohistioblast
(Stem Cell)

Myeloblast —— 2

Monoblast —— 3

Neutrophilic —— 9
Myelocyte

Eosinophilic —— 10
Myelocyte

Basophilic —— 11
Myelocyte

Megakaryocyte —— 12

Neutrophilic —— 14
Metamyelocyte

Eosinophilic —— 15
Metamyelocyte

Basophilic —— 16
Metamyelocyte

Mature
Basophil

23

Mature
Eosinophil

22

Mature
Neutrophil

21

Proerythroblast — 8

Normoblast — 13

Platelet

20

Reticulocyte — 18

19

Erythroblast
(Adult Red
Blood Cell)

No. NS16 ©1988 AMERICAN MAP CORP

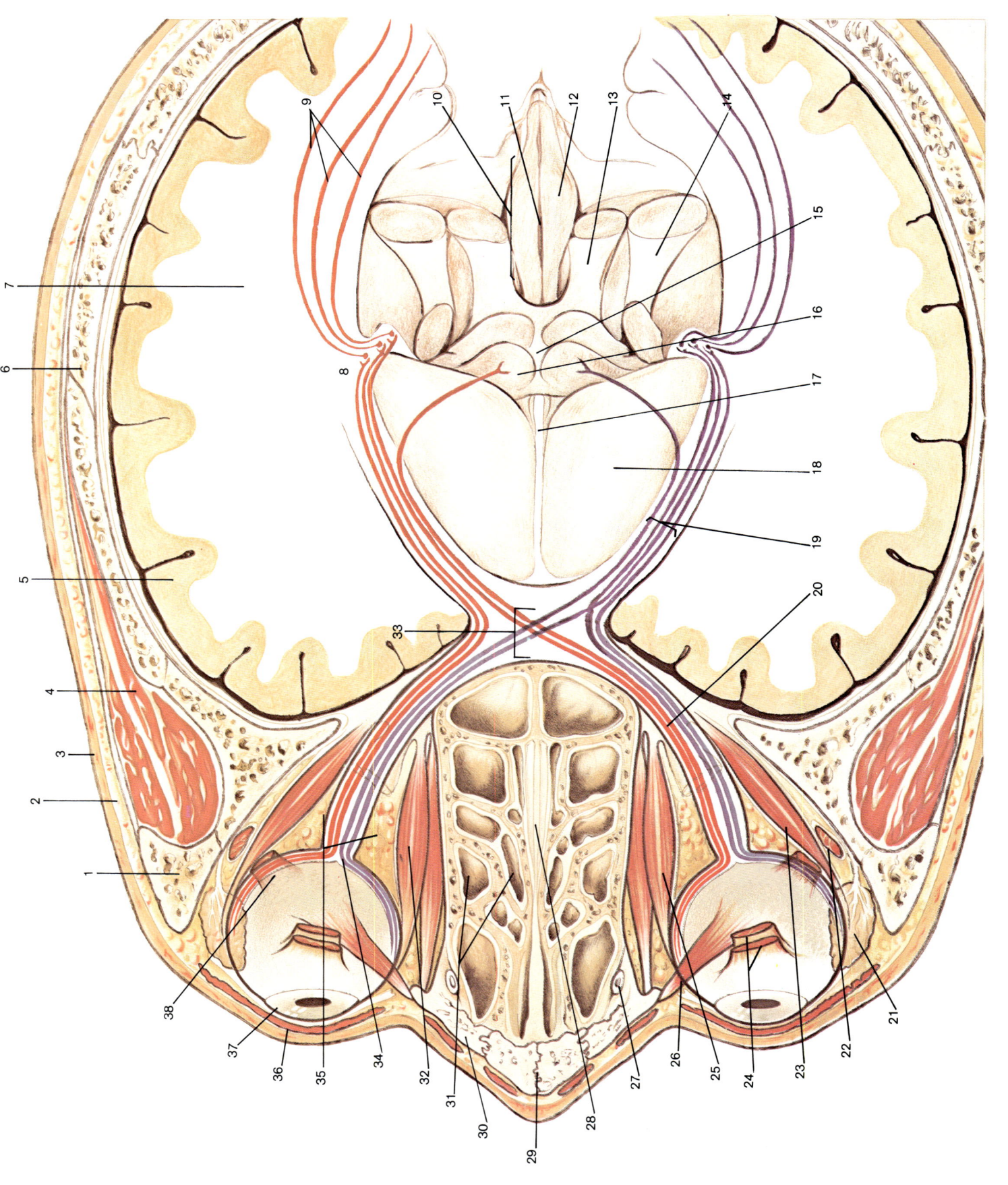

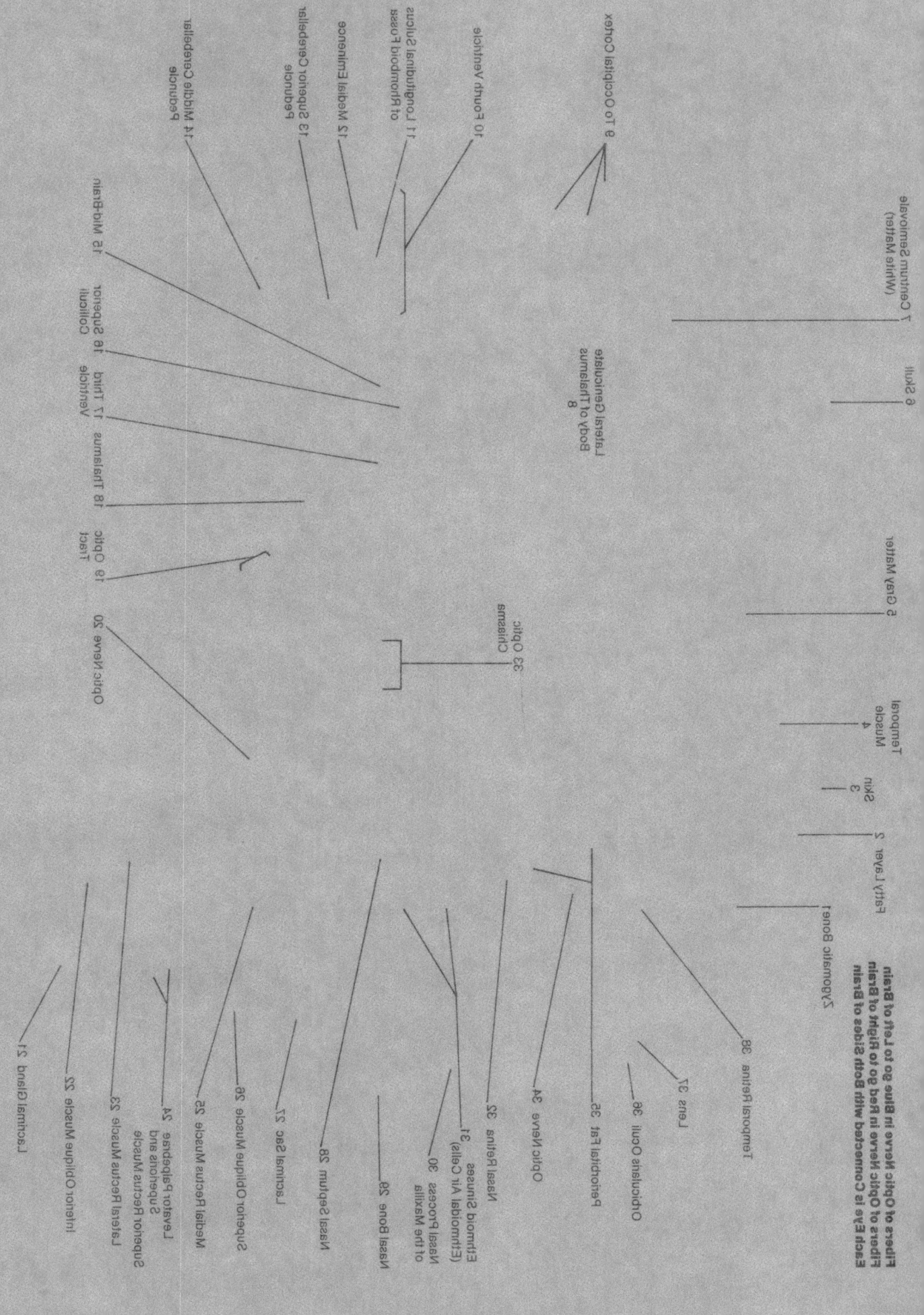

EYE – Sight as a Function of the Brain

7 Centrum Semiovale (White Matter)

6 Skull

5 Gray Matter

4 Temporal Muscle

3 Skin

2 Fatty Layer

1 Zygomatic Bone

9 To Occipital Cortex

33 Optic Chiasma

21 Lacrimal Gland

22 Inferior Oblique Muscle

23 Lateral Rectus Muscle

24 Superior Rectus Muscle and Levator Palpebrae Superioris

25 Medial Rectus Muscle

26 Superior Oblique Muscle

27 Lacrimal Sac

28 Nasal Septum

29 Nasal Bone

30 Nasal Process of the Maxilla

31 Ethmoidal Sinuses (Ethmoidal Air Cells)

32 Nasal Retina

34 Optic Nerve

35 Periorbital Fat

36 Orbicularis Oculi

37 Lens

38 Temporal Retina

Fibers of Optic Nerve go to Left of Brain in Blue
Fibers of Optic Nerve go to Right of Brain in Red
Each Eye is Connected with Both Sides of Brain

Fibers of Optic Nerve in Blue go to Left of Brain
Fibers of Optic Nerve in Red go to Right of Brain
Each Eye is Connected with Both Sides of Brain

6 Skull

7 Centrum Semiovale
(White Matter)

9 To Occipital Cortex

10 Fourth Ventricle

11 Longitudinal Sulcus
of Rhomboid Fossa

12 Medial Eminence

13 Superior Cerebellar
Peduncle

14 Middle Cerebellar
Peduncle

15 Mid-Brain

16 Superior
Colliculi

17 Third
Ventricle

18 Thalamus

19 Optic
Tract

20 Optic Nerve

Lateral Geniculate
Body of Thalamus
8

5 Gray Matter

33 Optic
Chiasma

Temporal
Muscle
4

Skin
3

Fatty Layer 2

Zygomatic Bone 1

Temporal Retina 38

Lens 37

Orbicularis Oculi 36

Periorbital Fat 35

Optic Nerve 34

Nasal Retina 32

Ethmoid Sinuses 31
(Ethmoidal Air Cells)

Nasal Process
of the Maxilla 30

Nasal Bone 29

Nasal Septum 28

Lacrimal Sac 27

Superior Oblique Muscle 26

Medial Rectus Muscle 25

Levator Palpebrae 24
Superioris and
Superior Rectus Muscle

Lateral Rectus Muscle 23

Inferior Oblique Muscle 22

Lacrimal Gland 21

EYE – Sight as a Function of the Brain

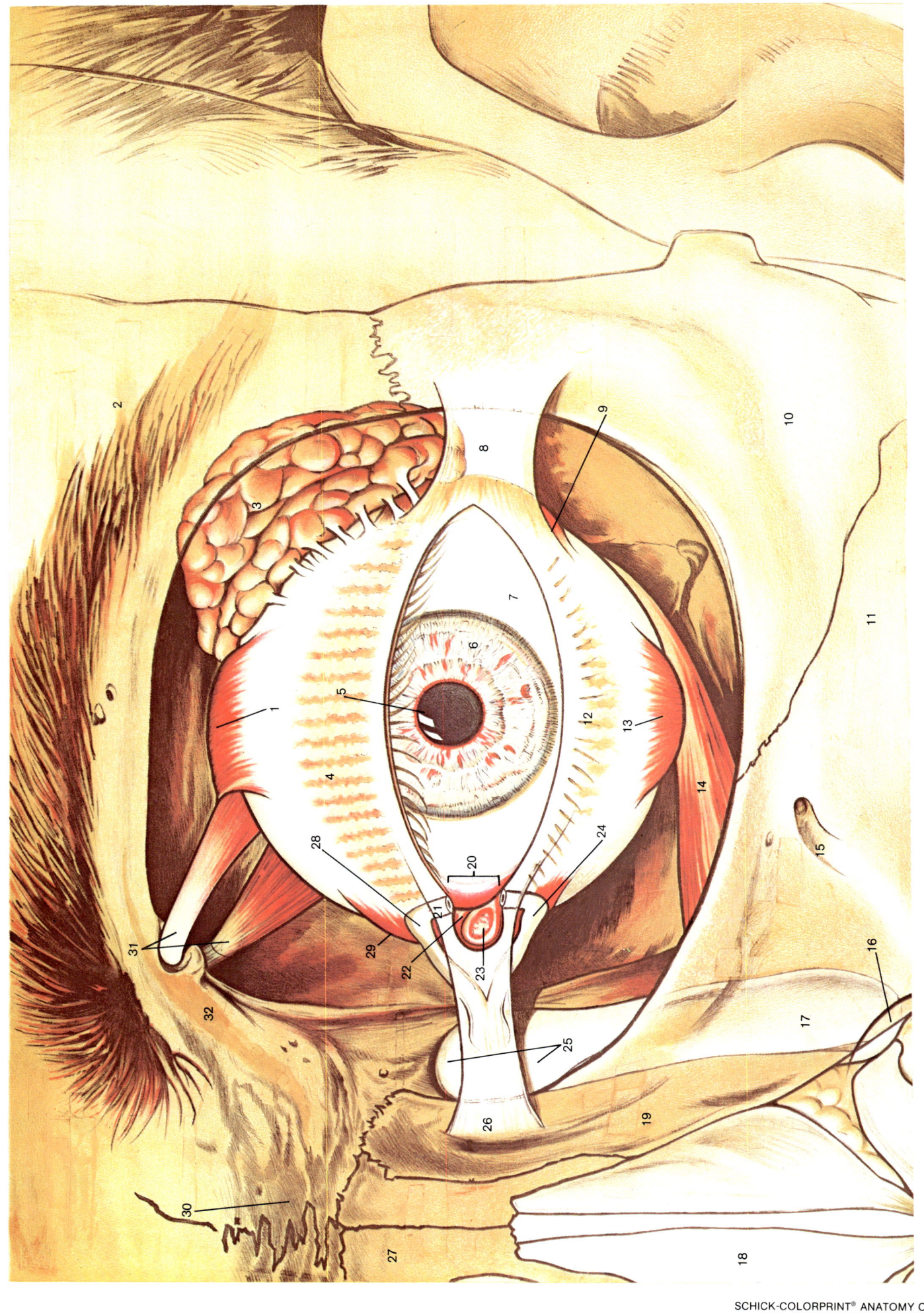

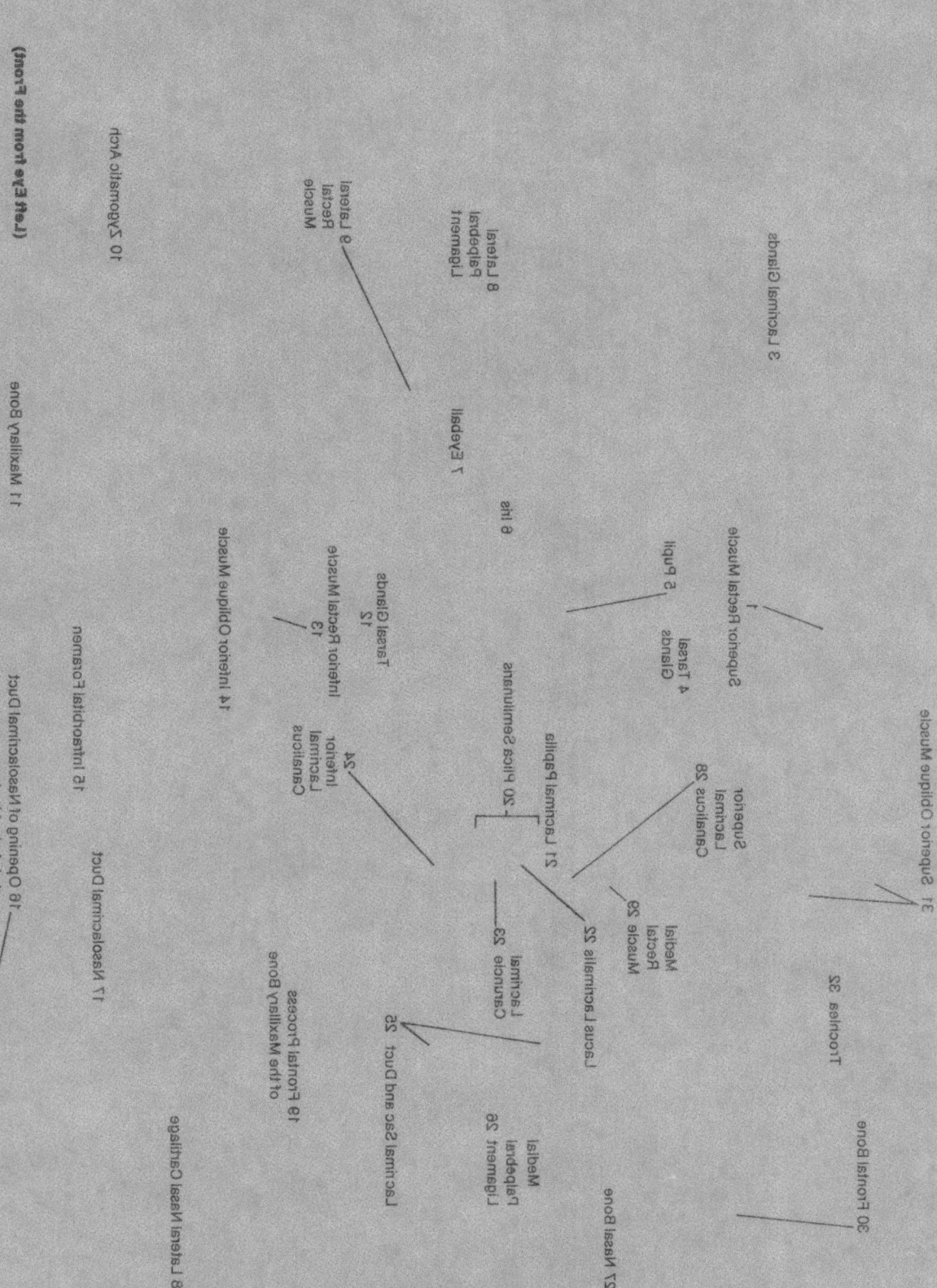

2 Frontal Bone

3 Lacrimal Glands

1 Superior Rectal Muscle

4 Tarsal Glands

5 Pupil

6 Iris

7 Eyeball

8 Palpebral Ligament

9 Lateral Rectal Muscle

10 Zygomatic Arch

11 Maxillary Bone

12 Tarsal Glands

13 Inferior Rectal Muscle

14 Inferior Oblique Muscle

15 Infraorbital Foramen

16 Opening of Nasolacrimal Duct in Inferior Meatus

17 Nasolacrimal Duct

18 Lateral Nasal Cartilage

19 Frontal Process of the Maxillary Bone

20 Lacrimal Papilla

21 Lacrimal Puncta Semilunaris

22 Lacrimal Caruncle

23 Lacrimal Caruncle

24 Inferior Lacrimal Canaliculus

25 Lacrimal Sac and Duct

26 Medial Palpebral Ligament

27 Nasal Bone

28 Superior Lacrimal Canaliculus

29 Medial Rectal Muscle

30 Frontal Bone

31 Superior Oblique Muscle

32 Trochlea

(Left Eye from the Front)

EYE – Protective Mechanism

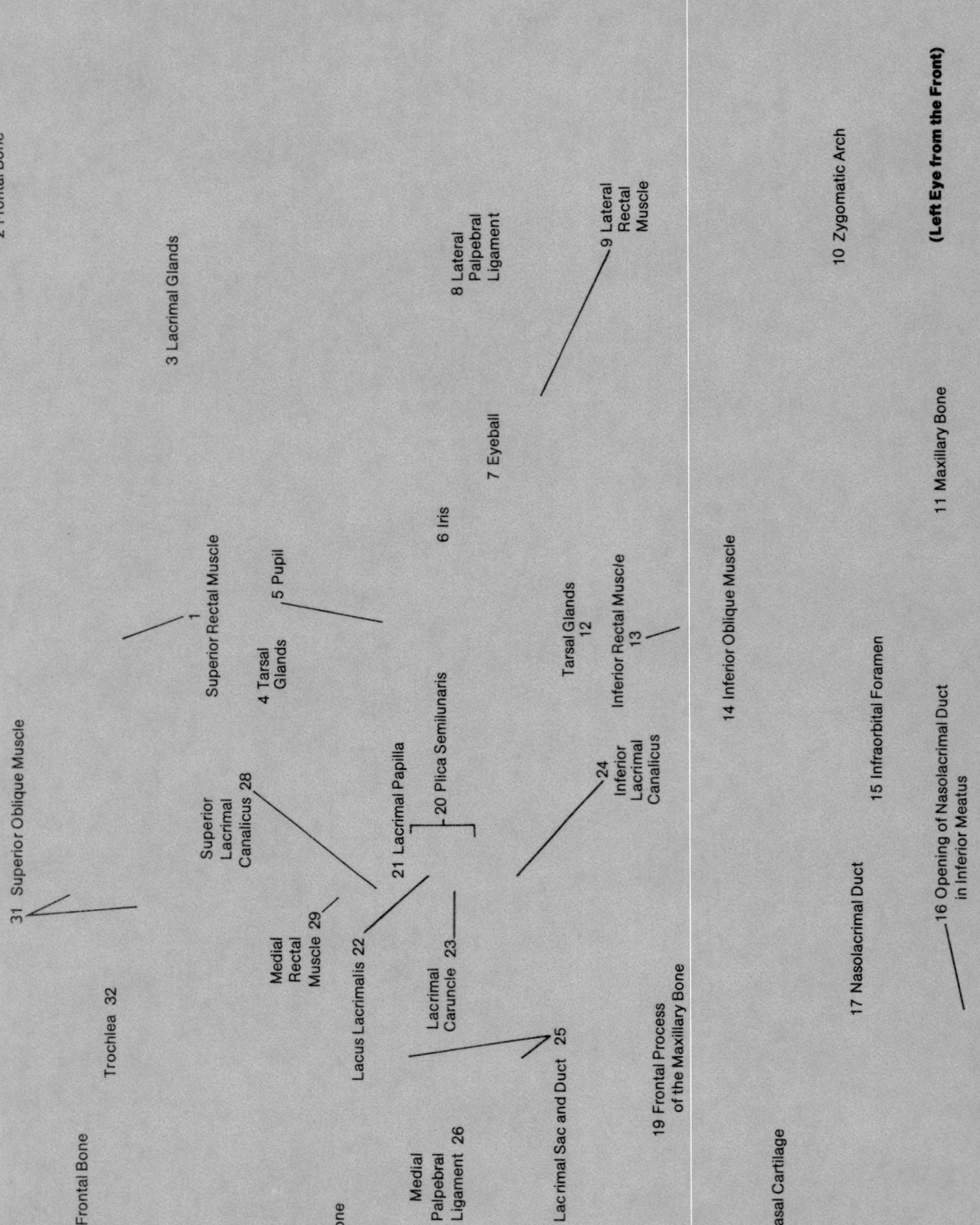

2 Frontal Bone

3 Lacrimal Glands

1

Superior Rectal Muscle

4 Tarsal Glands

5 Pupil

6 Iris

7 Eyeball

8 Lateral Palpebral Ligament

9 Lateral Rectal Muscle

10 Zygomatic Arch

11 Maxillary Bone

12 Tarsal Glands

13 Inferior Rectal Muscle

14 Inferior Oblique Muscle

15 Infraorbital Foramen

16 Opening of Nasolacrimal Duct in Inferior Meatus

17 Nasolacrimal Duct

18 Lateral Nasal Cartilage

19 Frontal Process of the Maxillary Bone

20 Plica Semilunaris

21 Lacrimal Papilla

22 Lacus Lacrimalis

23 Lacrimal Caruncle

24 Inferior Lacrimal Canaliculus

25 Lacrimal Sac and Duct

26 Medial Palpebral Ligament

27 Nasal Bone

28 Superior Lacrimal Canaliicus

29 Medial Rectal Muscle

30 Frontal Bone

31 Superior Oblique Muscle

32 Trochlea

(Left Eye from the Front)

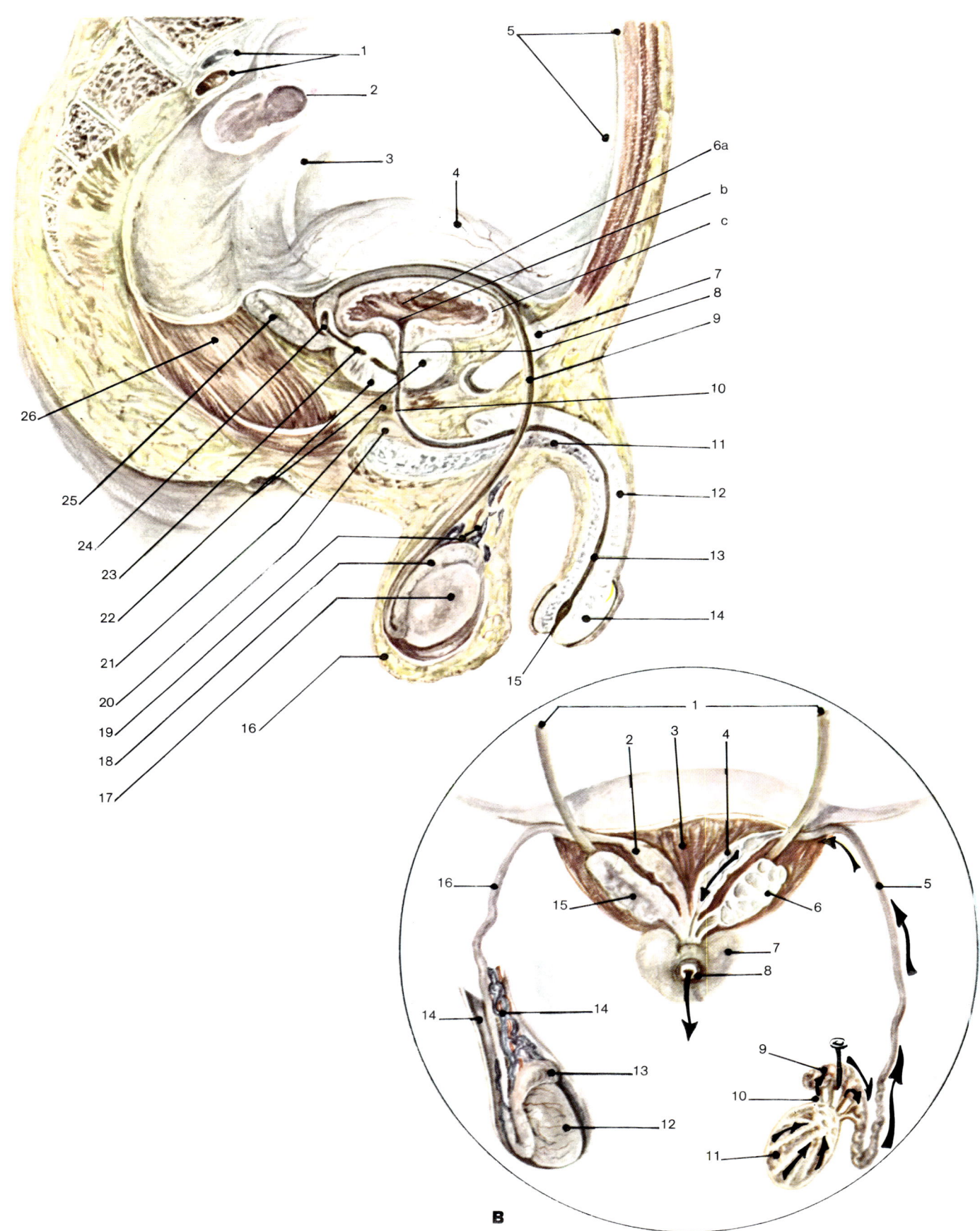

A

1
2
3
4
5
6a
b
c
7
8
9
10
11
12
13
14
15
16
17
18
19
20
21
22
23
24
25
26

B

1
2
3
4
5
6
7
8
9
10
11
12
13
14
15
16

SCHICK-COLORPRINT® ANATOMY CHART
MALE REPRO. ORGANS
No. NS13 © 1988 AMERICAN MAP CORP.

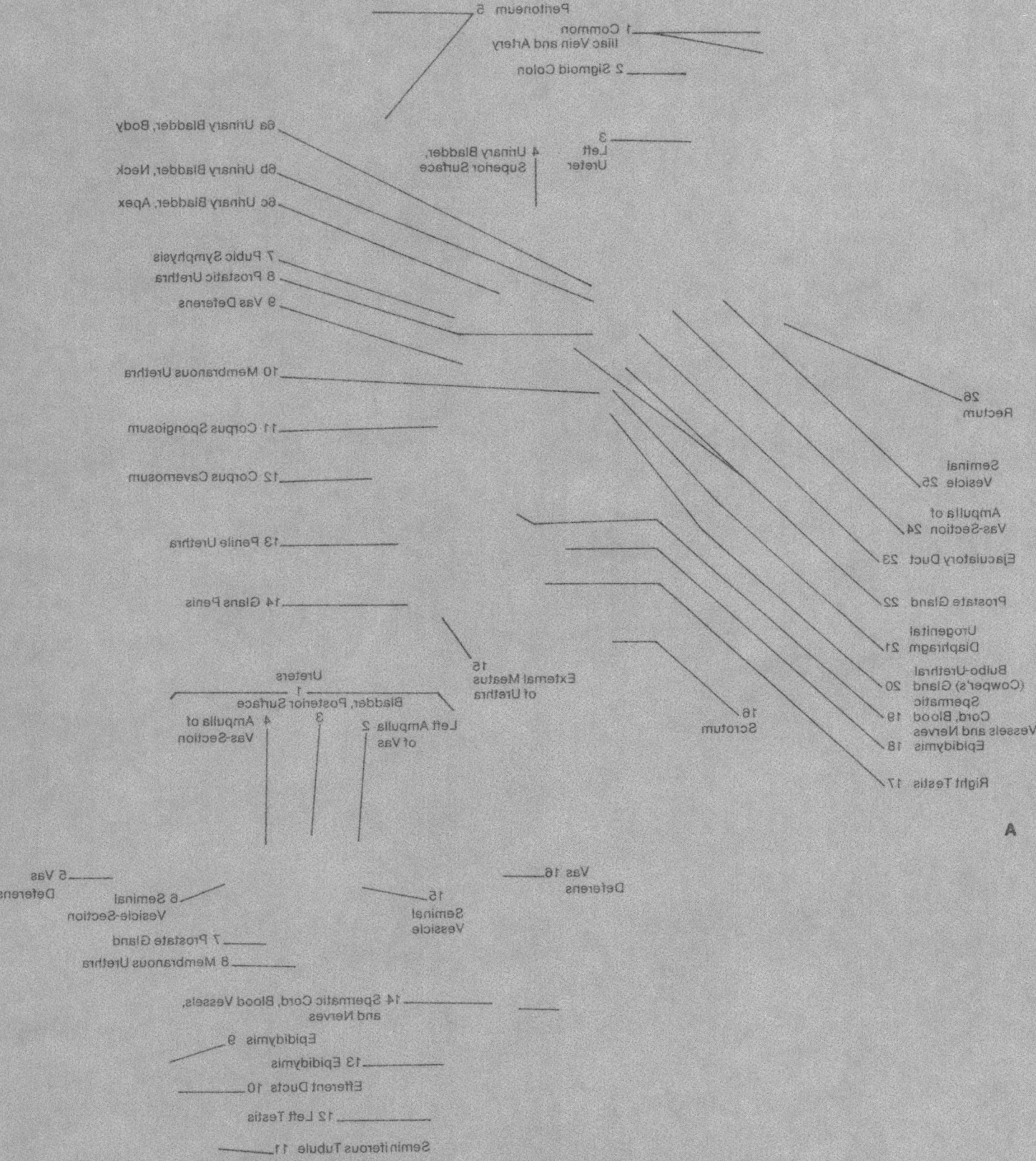

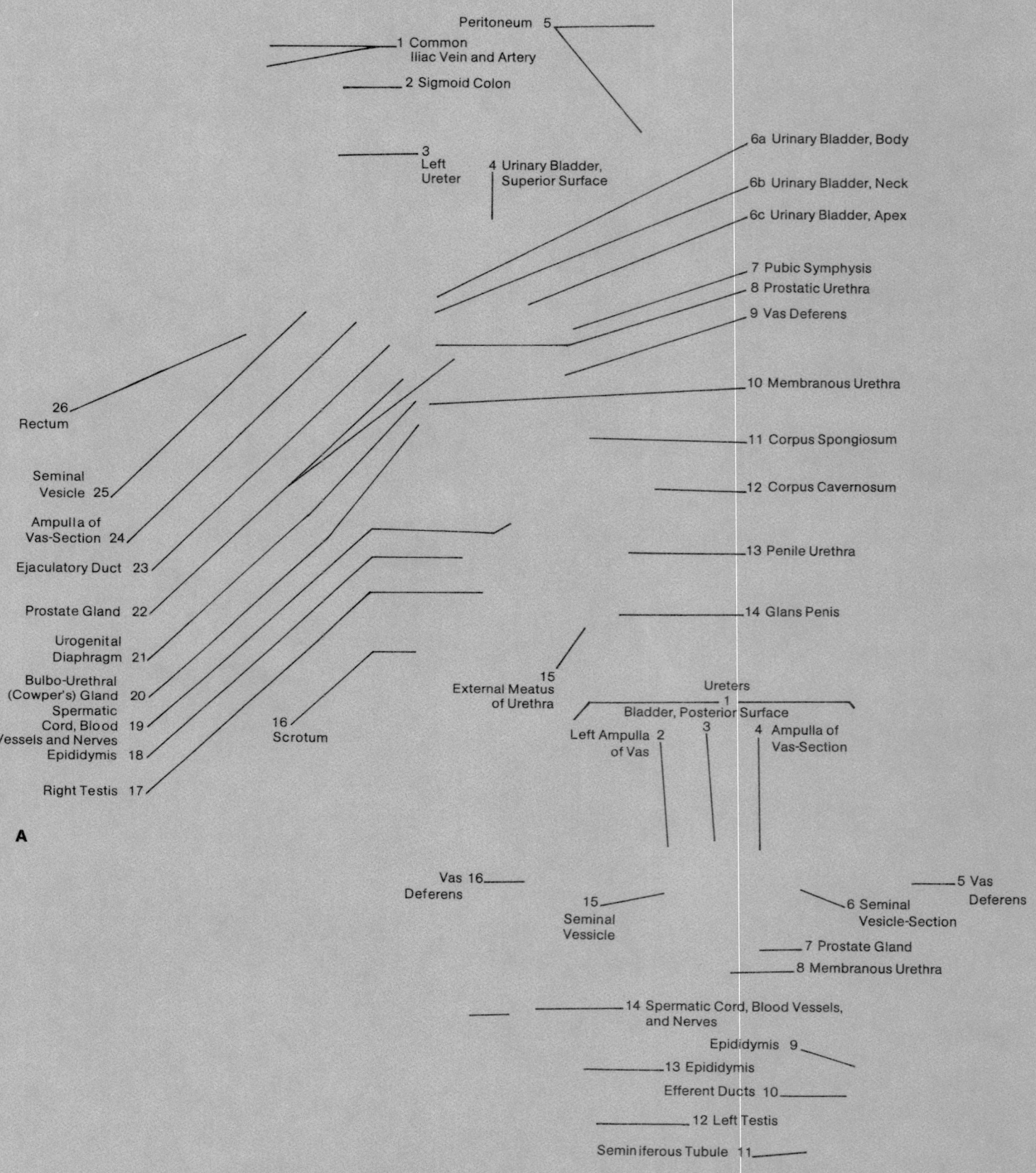

Peritoneum 5

1 Common Iliac Vein and Artery

2 Sigmoid Colon

3 Left Ureter

4 Urinary Bladder, Superior Surface

6a Urinary Bladder, Body

6b Urinary Bladder, Neck

6c Urinary Bladder, Apex

7 Pubic Symphysis

8 Prostatic Urethra

9 Vas Deferens

10 Membranous Urethra

11 Corpus Spongiosum

12 Corpus Cavernosum

13 Penile Urethra

14 Glans Penis

26 Rectum

Seminal Vesicle 25

Ampulla of Vas-Section 24

Ejaculatory Duct 23

Prostate Gland 22

Urogenital Diaphragm 21

Bulbo-Urethral (Cowper's) Gland 20

Spermatic Cord, Blood Vessels and Nerves 19

Epididymis 18

Right Testis 17

16 Scrotum

15 External Meatus of Urethra

A

Ureters

1 Bladder, Posterior Surface

Left Ampulla 2 of Vas

3

4 Ampulla of Vas-Section

Vas 16 Deferens

15 Seminal Vessicle

5 Vas Deferens

6 Seminal Vesicle-Section

7 Prostate Gland

8 Membranous Urethra

14 Spermatic Cord, Blood Vessels, and Nerves

Epididymis 9

13 Epididymis

Efferent Ducts 10

12 Left Testis

Seminiferous Tubule 11

B

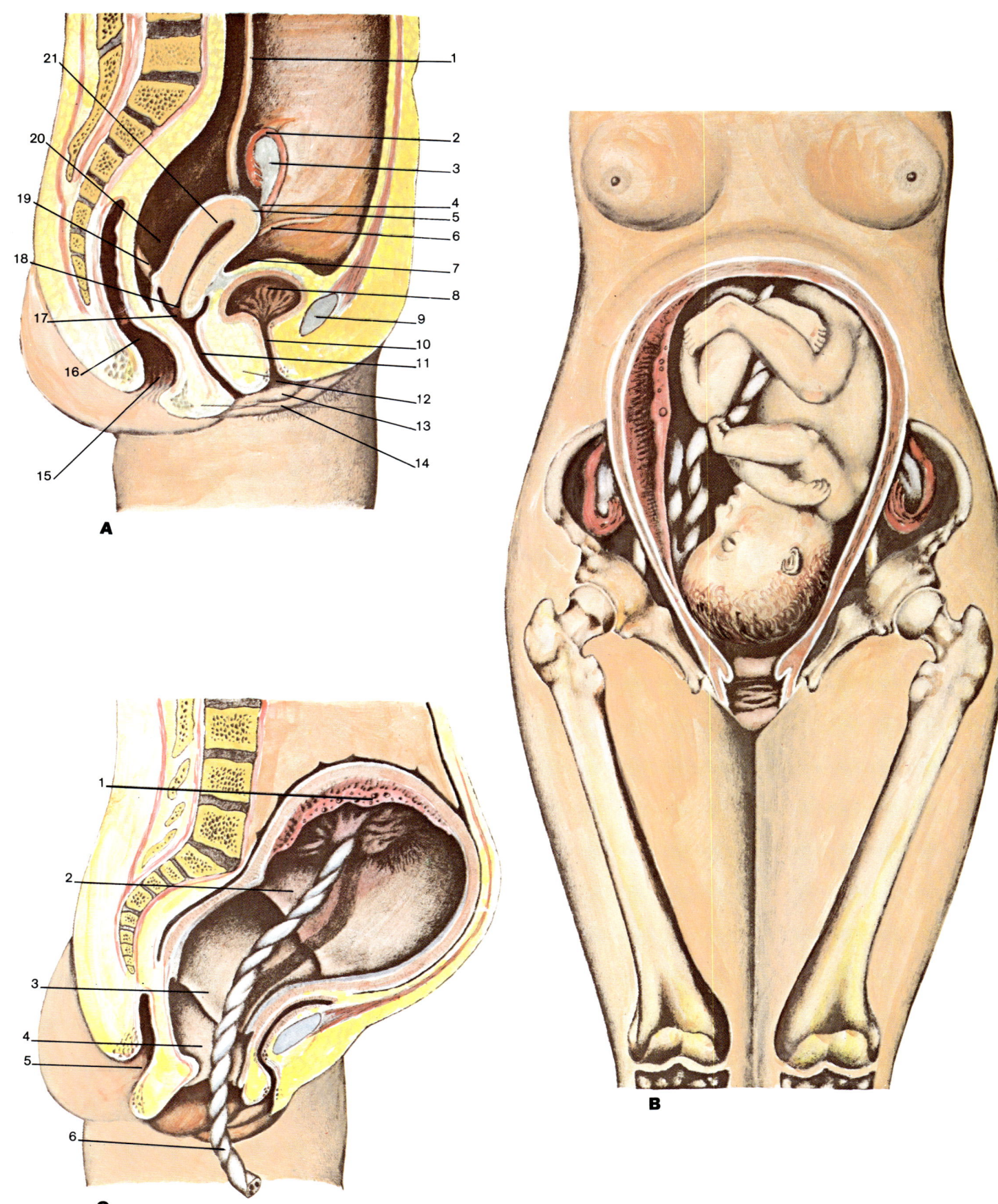

SCHICK-COLORPRINT® ANATOMY CHART
FEMALE REPRO. ORGANS AND PREGNANCY AT TERM
No. NS12 © 1988 AMERICAN MAP CORP.

Corpus
of Uterus 21

Rectouterine
Pouch
(of Douglas)
20

1 Ureter

Sacrouterine
Ligament
19

2 Oviduct

3 Ovary

Uterine
Orifice
18

4 Fundus of Uterus
5 Uterus

6 Round Ligament

7 Vesiculouterine Pouch

3

Vagina

FEMALE REPRODUCTIVE ORGANS AND PREGNANCY AT TERM

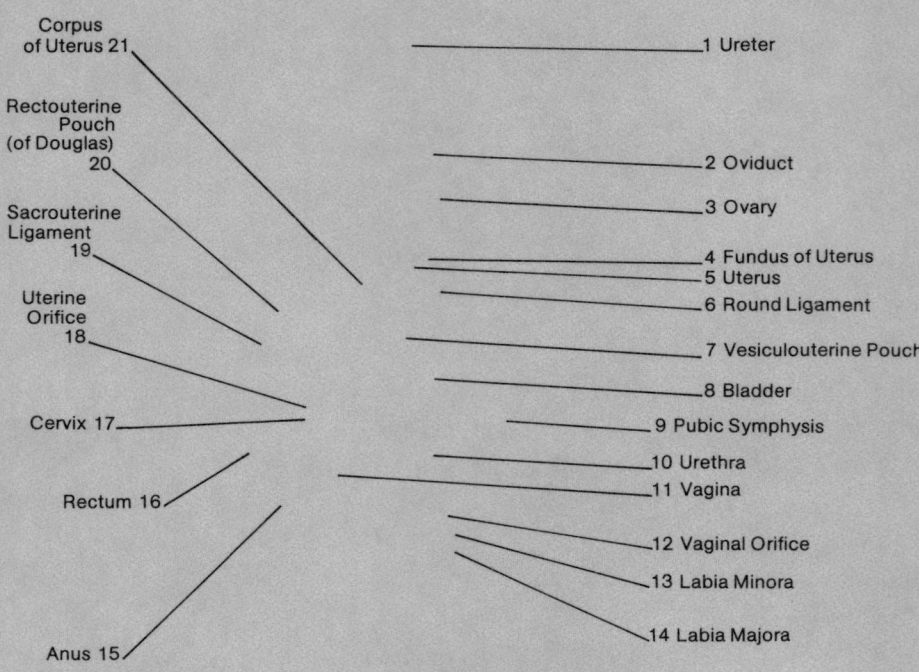

Corpus
of Uterus 21

Rectouterine
Pouch
(of Douglas)
20

Sacrouterine
Ligament
19

Uterine
Orifice
18

Cervix 17

Rectum 16

Anus 15

1 Ureter

2 Oviduct

3 Ovary

4 Fundus of Uterus
5 Uterus
6 Round Ligament

7 Vesiculouterine Pouch

8 Bladder

9 Pubic Symphysis

10 Urethra
11 Vagina

12 Vaginal Orifice

13 Labia Minora

14 Labia Majora

**A. Section Through the Pelvis
of a Non-Pregnant Woman**

Placenta 1

Internal Os
of Cervix 2

External Os
of Cervix
(Uterine
Orifice)
3

Vagina
4

Anus 5

Umbilical Cord
6

B. Pregnancy at Term

**C. Section Through the Uterus
of a Pregnant Woman After Delivery**

No. NS12 © 1988 AMERICAN MAP CORP.

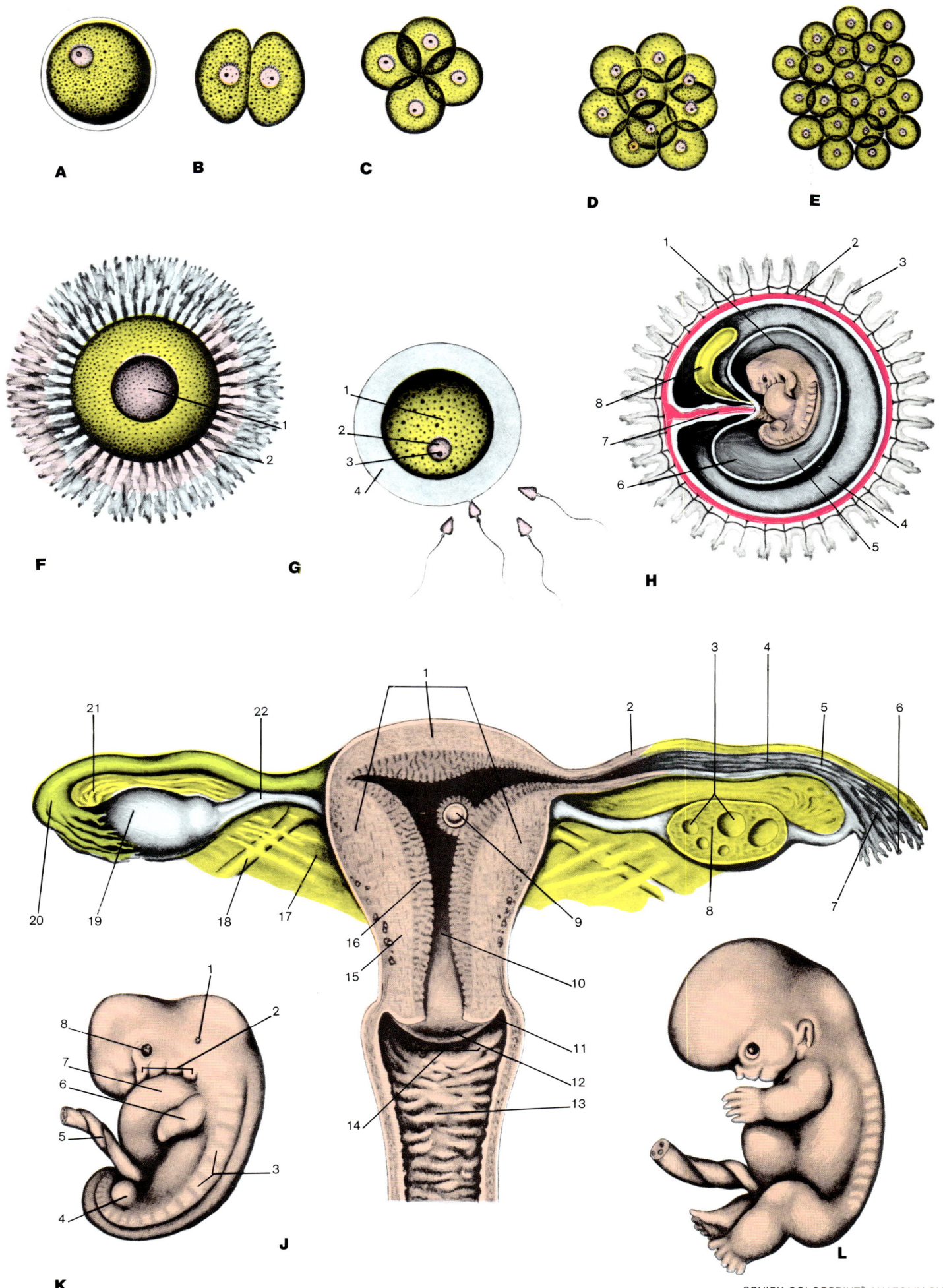

A

B

C

D

E

F

G

H

1 2 3

8

7

6

5

4

3 4

2

5

6

21 22 1

20 19 18 17 16 15 14

9 10 11 12 13

8 7

K

1 2

8

7

6

5

4

3

J

L

SCHICK-COLORPRINT® ANATOMY CHART
DEVELOPMENT OF THE HUMAN EMBRYO
No. NS11 © 1988 AMERICAN MAP CORP.

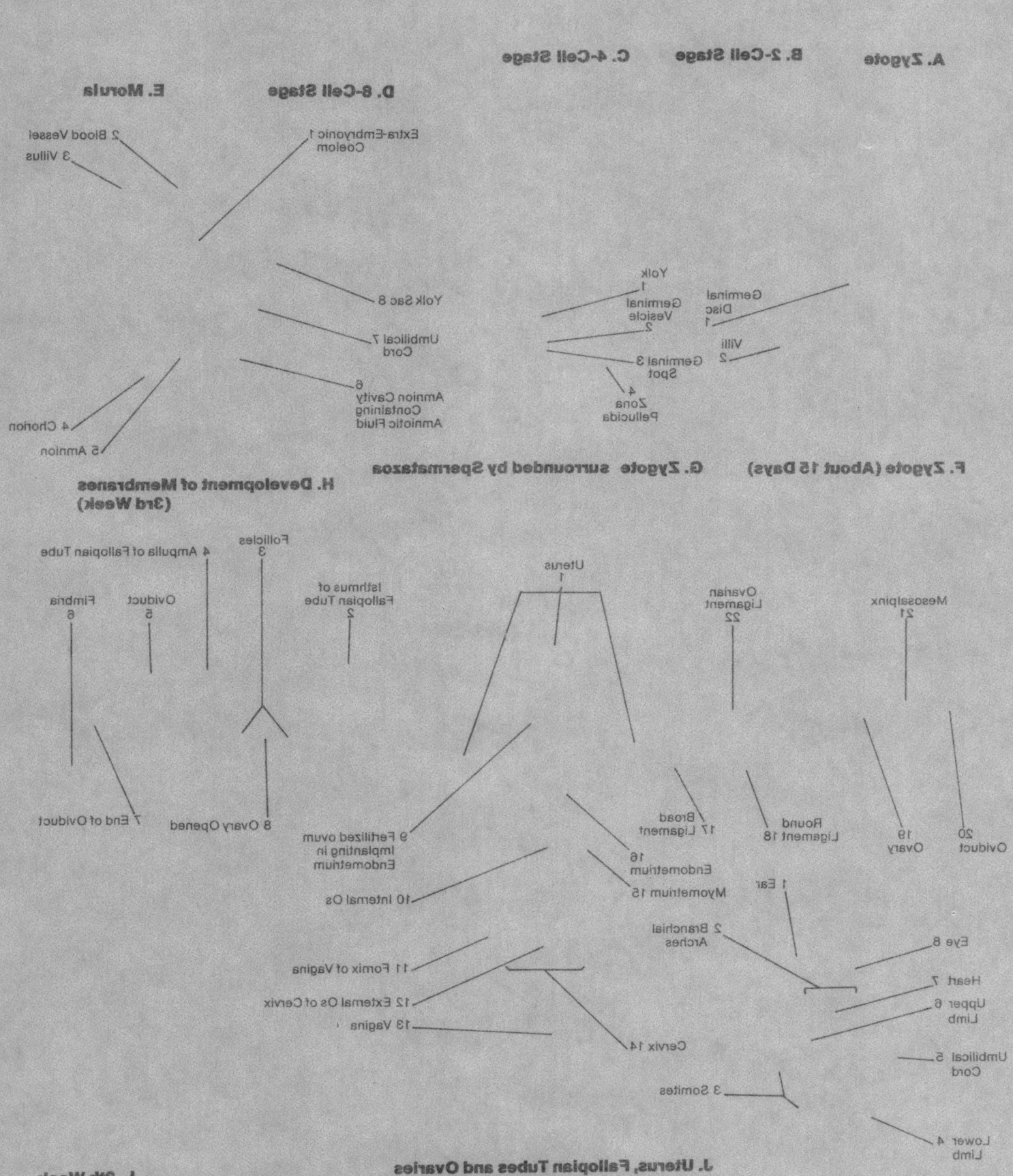

DEVELOPMENT OF THE HUMAN EMBRYO

DEVELOPMENT OF THE HUMAN EMBRYO

A. Zygote   **B. 2-Cell Stage**   **C. 4-Cell Stage**

D. 8-Cell Stage   **E. Morula**

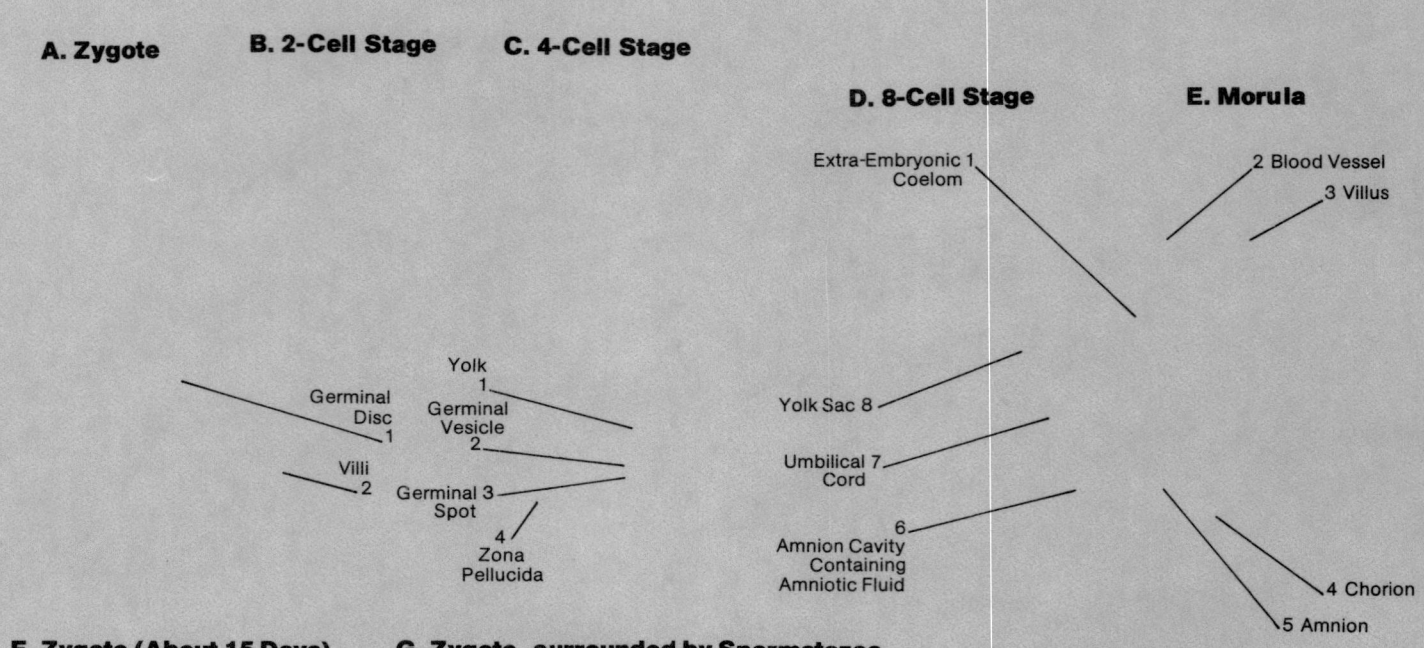

Germinal Disc 1
Villi 2

Yolk 1
Germinal Vesicle 2
Germinal Spot 3
4
Zona Pellucida

Extra-Embryonic 1 Coelom
Yolk Sac 8
Umbilical 7 Cord
6
Amnion Cavity Containing Amniotic Fluid

2 Blood Vessel
3 Villus
4 Chorion
5 Amnion

F. Zygote (About 15 Days)   **G. Zygote surrounded by Spermatazoa**

H. Development of Membranes (3rd Week)

Mesosalpinx 21
Ovarian Ligament 22
Uterus 1
Isthmus of Fallopian Tube 2

Follicles 3
4 Ampulla of Fallopian Tube
Oviduct 5
Fimbria 6

20 Oviduct
19 Ovary
Round Ligament 18
17 Broad Ligament
16
Endometrium
Myometrium 15

9 Fertilized ovum Implanting in Endometrium
10 Internal Os

8 Ovary Opened
7 End of Oviduct

1 Ear
2 Branchial Arches
Eye 8
Heart 7
Upper 6 Limb
Umbilical 5 Cord

11 Fornix of Vagina
12 External Os of Cervix
13 Vagina
Cervix 14
3 Somites

Lower 4 Limb

J. Uterus, Fallopian Tubes and Ovaries

L. 8th Week

K. Embryo (5th Week)

No. NS11 © 1988 AMERICAN MAP CORP.

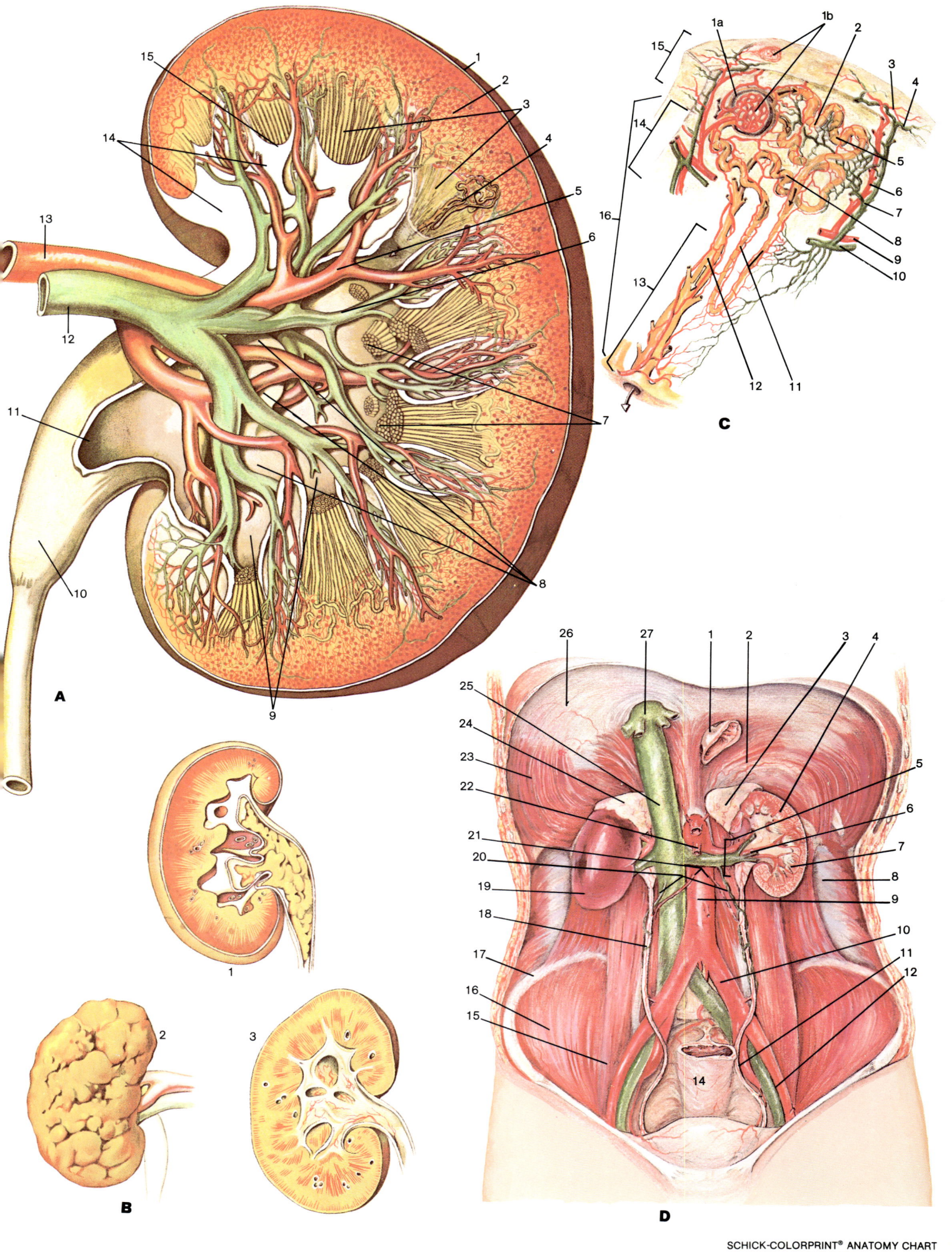

SCHICK-COLORPRINT® ANATOMY CHART
KIDNEYS
No. NS10 © 1988 AMERICAN MAP CORP.

A. Longitudinal Section

1. Capsule
2. Cortex
3. Medulla
4. Nephron
5. Ramification of Arteries
6. Ramification of Veins
7. Pyramids
8. Major Calyces
9. Minor Calyces
10. Ureter
11. Renal Pelvis
12. Renal Vein
13. Renal Artery
14. Renal Sinus
15. Papilla of Pyramid

C. Nephron

1. Renal Corpuscle
   a. Bowman's Capsule
   b. Glomeruli
2. Proximal Convoluted Tubule
3. Perforating Artery
4. Stellate Vein
5. Renal Tubule
6. Interlobular Artery
7. Interlobular Vein
8. Distal Convoluted Tubule
9. Arcuate Artery
10. Arcuate Vein
11. Henle's Loop
12. Collecting Tubule
13. Inner Zone
14. Outer Zone
15. Cortex
16. Medulla

B. Kidney Disorders

Staghorn Calculus 1
(Kidney Stones)

Pyelonephritis
(Cross Section)

2 Alcoholic
3
Kidney

11. Internal Iliac Artery
27. Hepatic Veins

KIDNEYS

A. Longitudinal Section

1. Capsule
2. Cortex
3. Medulla
4. Nephron
5. Ramification of Arteries
6. Ramification of Veins
7. Pyramids
8. Major Calyces
9. Minor Calyces
10. Ureter
11. Renal Pelvis
12. Renal Vein
13. Renal Artery
14. Renal Sinus
15. Papilla of Pyramid

C. Nephron

1. Renal Corpuscle
   a. Bowman's Capsule
   b. Glomeruli
2. Proximal Convoluted Tubule
3. Perforating Artery
4. Stellate Vein
5. Renal Tubule
6. Interlobular Artery
7. Interlobular Vein
8. Distal Convoluted Tubule
9. Arcuate Artery
10. Arcuate Vein
11. Henle's Loop
12. Collecting Tubule
13. Inner Zone
14. Outer Zone
15. Cortex
16. Medulla

B. Kidney Disorders

Staghorn Calculus 1
(Kidney Stones)

Pyelonephritis
(Cross Section)

2 Alcholic   3
Kidney

D. Location

1. Esophagus
2. Cortex
3. Suprarenal (Adrenal) Gland
4. Left Kidney Section
5. Hilum of Kidney
6. Renal Pelvis and Blood Vessels
7. Medulla (Pyramids)
8. Abdominal Wall
9. Abdominal Aorta
10. Common Iliac Artery
11. Internal Iliac Artery
12. External Iliac Artery
13. Urinary Bladder
14. Rectum
15. Psoas Major Muscle
16. Iliacus Muscle
17. Iliac Crest
18. Ureter
19. Right Kidney
20. Gonadal Vein
21. Gonadal Artery
22. Renal Vein
23. 11th Thoracic Rib
24. Suprarenal (Adrenal) Gland
25. Inferior Vena Cava
26. Diaphragm
27. Hepatic Veins

No. NS10 © 1988 AMERICAN MAP CORP.

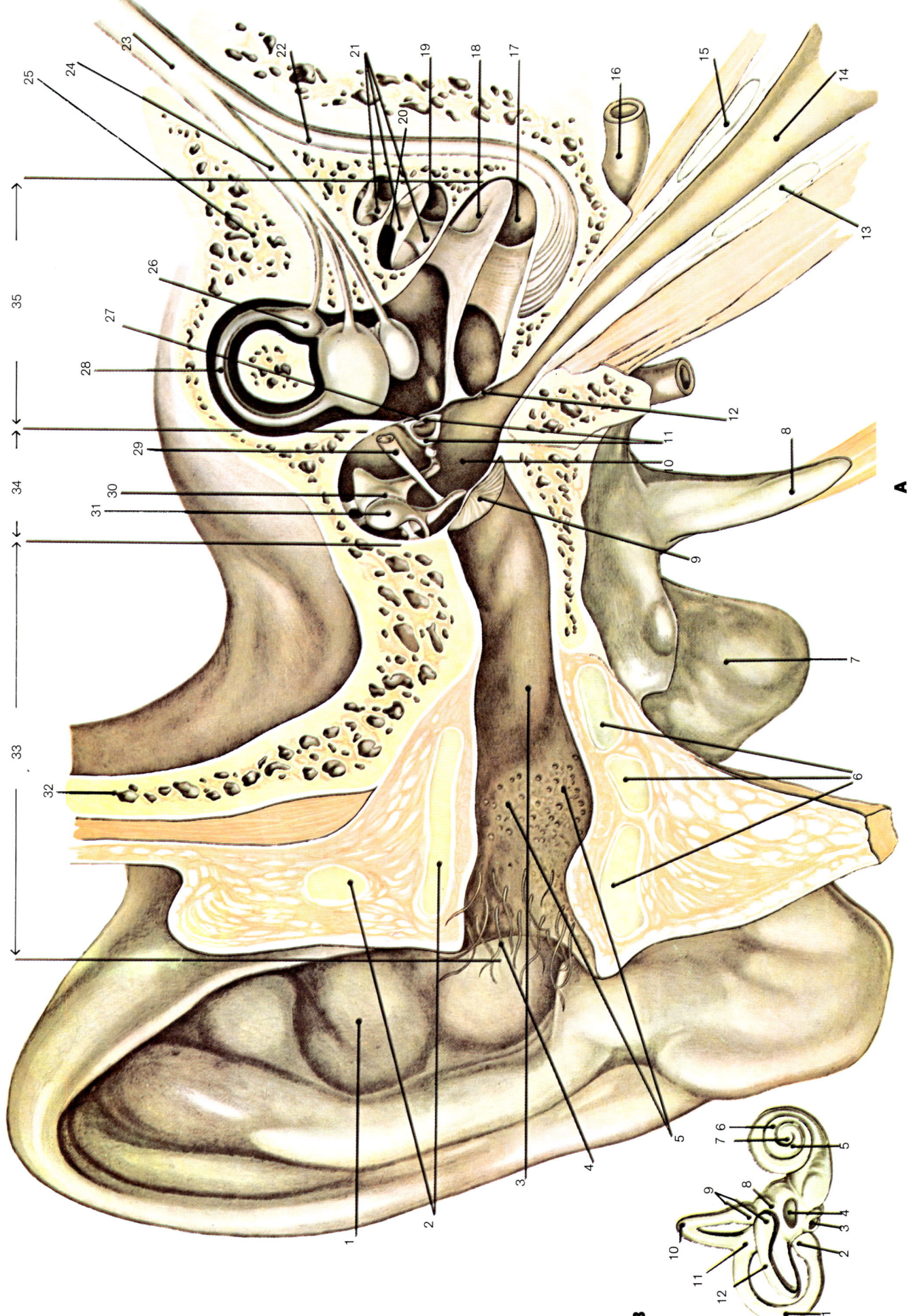

SCHICK-COLORPRINT® ANATOMY CHART
EAR
No. NS9 © 1988 AMERICAN MAP CORP.

EAR

A. Ear (Sectioned Vertically)

Inner Ear (Labyrinth) 35 | Middle Ear 34 | External Ear 33

- Petrous Portion of Temporal Bone 25
- 24 Vestibular Nerve
- 23 Auditory Nerve
- 22 Cochlear Nerve
- 21 Scala Vestibuli
- 20 Modiolus
- 19 Scala Tympani
- 18 Spiral Lamina
- 17 Cochlea
- 26 Ampulla
- 27 Oval Window
- 28 Semicircular Canal
- 29 Tensor Tympani (Cut) Muscle
- 30 Incus
- 31 Malleus
- 16 Internal Carotid Artery
- 15 Cartilage
- 14 Eustachian Tube
- 13 Cartilage
- 7 Mastoid Process of Temporal Bone
- 8 Styloid Process of Temporal Bone
- 9 Tympanic Membrane
- 10 Middle Ear
- 11 Stapes and Vestibular Fenestra
- 12 Round Window
- 32 Temporal Bone
- 32 External Ear

B. Right Bony Labyrinth

- 1. Posterior Duct
- 2. Ampulla
- 3. Round Window
- 4. Oval Window
- 5. Apical Spiral
- 6. Middle Spiral
- 7. Cupula
- 8. Vestibule
- 9. Ampulla
- 10. Superior Duct
- 11. Common Duct
- 12. Lateral Duct

- 1 External Ear (Auricle)
- 2 Cartilage
- 3 Medial Acoustic Meatus
- 4 Hairs
- 5 Cerumen Glands
- 6 Cartilage

EAR

Petrous Portion of Temporal Bone 25
24 Vestibular Nerve
23 Auditory Nerve
22 Cochlear Nerve
21 Scala Vestibuli
20 Modiolus
19 Scala Tympani
18 Spiral Lamina
17 Cochlea
16 Internal Carotid Artery
15 Cartilage
14 Eustachian Tube
13 Cartilage

Inner Ear (Labyrinth) 35
26 Ampulla
27 Oval Window
28 Semicircular Canal
29 Tensor Tympani Muscle (Cut)
12 Round Window
11 Stapes and Vestibular Fenestra
8 Styloid Process of Temporal Bone

Middle Ear 34
31 Malleus
30 Incus
Middle Ear 10
9 Tympanic Membrane
7 Mastoid Process of Temporal Bone

A. Ear (Sectioned Vertically)

External Ear 33
32 Temporal Bone
6 Cartilage

External Ear 1 (Auricle)
Cartilage 2
External Acoustic 3 Meatus
Hairs 4
5 Cerumen Glands

B. Right Bony Labyrinth

1. Posterior Duct
2. Ampulla
3. Round Window
4. Oval Window
5. Apical Spiral
6. Middle Spiral
7. Cupula
8. Vestibule
9. Ampulla
10. Superior Duct
11. Common Canal
12. Lateral Duct

No. NS9 © 1988 AMERICAN MAP CORP.

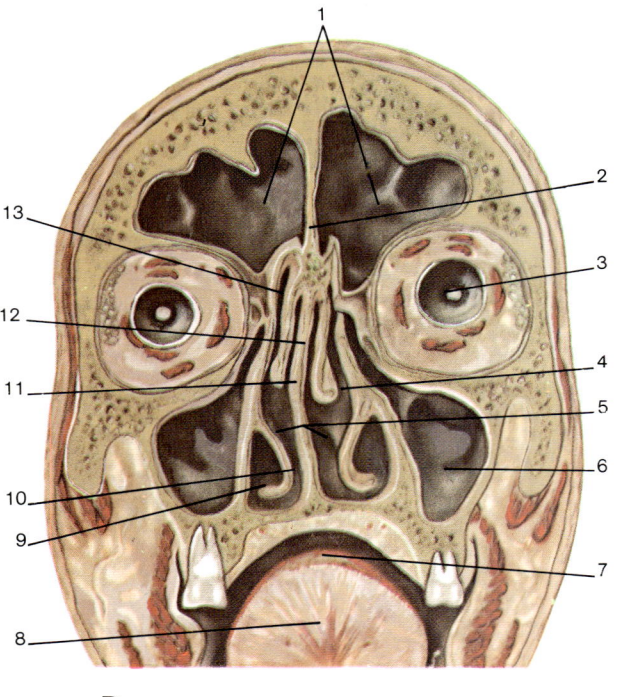

B

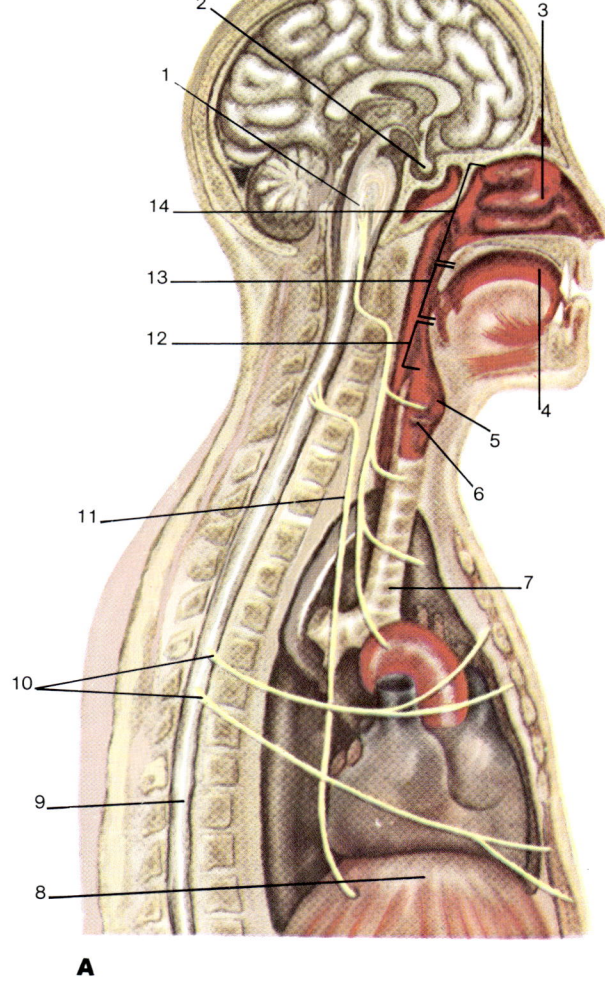

A

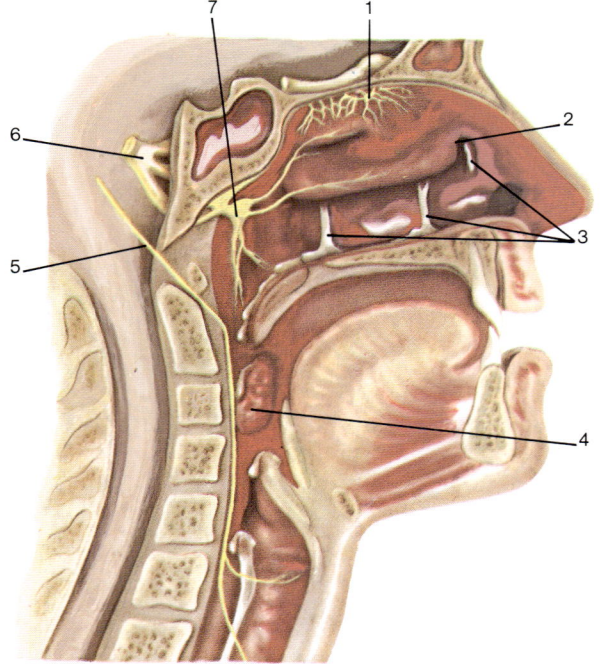

D

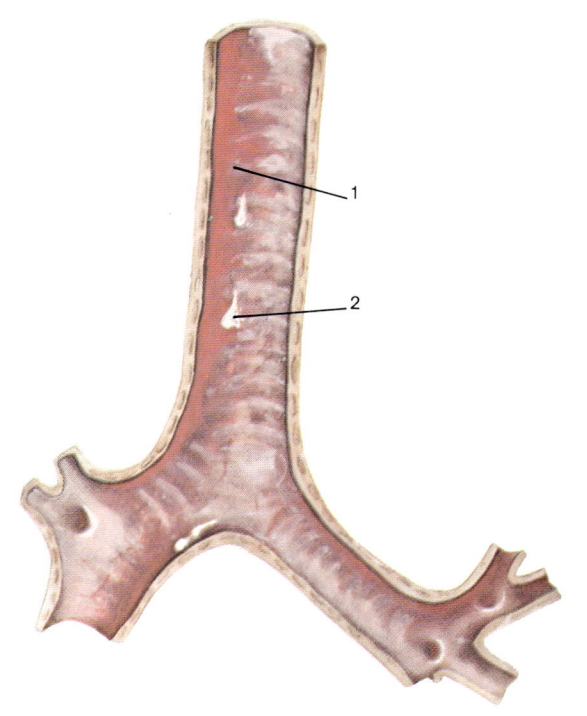

E

SCHICK-COLORPRINT® ANATOMY CHART
NERVOUS CONTROL OF RESPIRATION AND SYMPTOMS OF INFECTION
No. NS8 © 1988 AMERICAN MAP CORP.

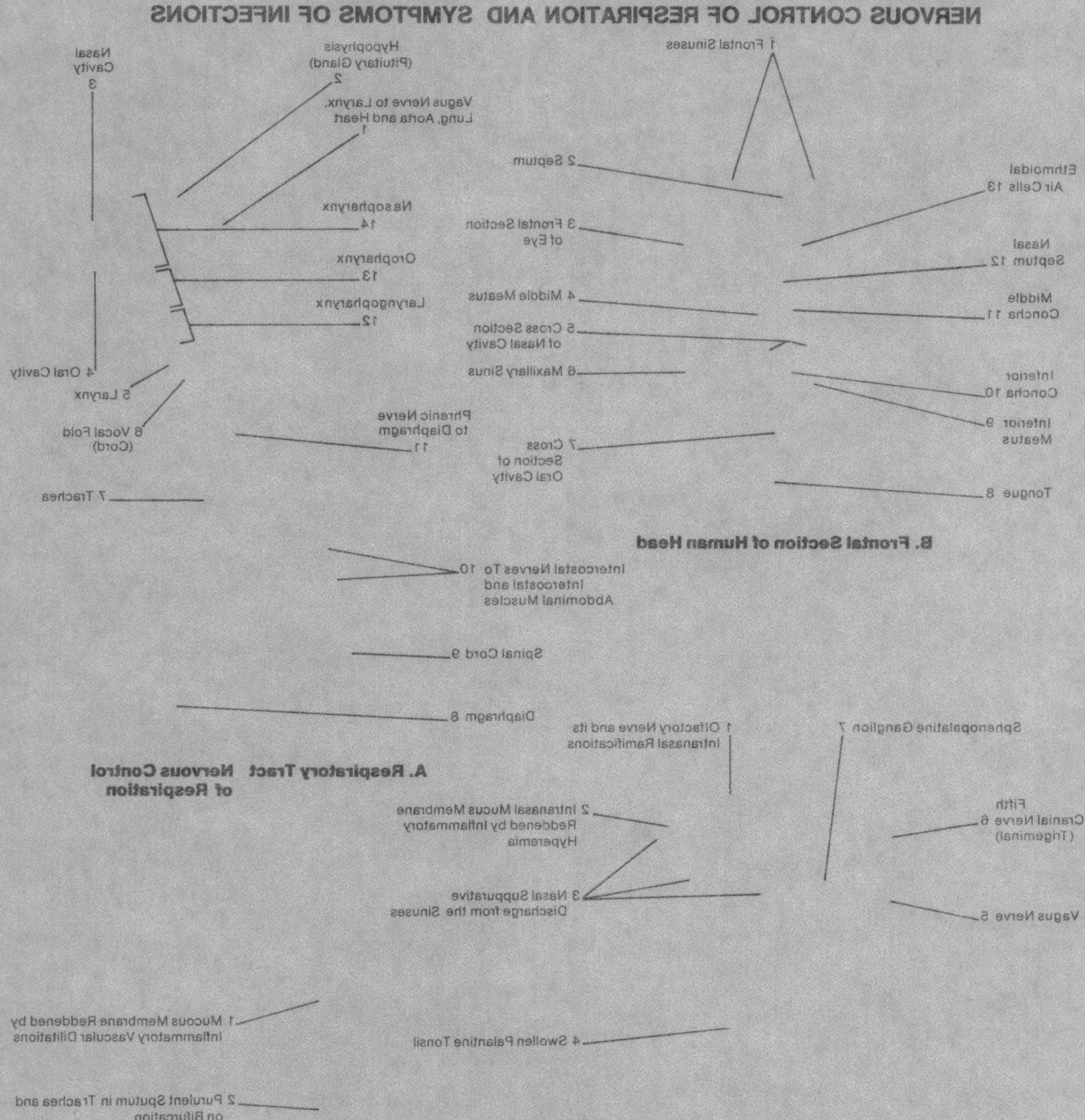

NERVOUS CONTROL OF RESPIRATION AND SYMPTOMS OF INFECTIONS

NERVOUS CONTROL OF RESPIRATION AND SYMPTOMS OF INFECTIONS

1 Frontal Sinuses

Hypophysis
(Pituitary Gland)
2

Nasal
Cavity
3

Vagus Nerve to Larynx,
Lung, Aorta and Heart
1

2 Septum

Ethmoidal
Air Cells 13

Nasopharynx
14

3 Frontal Section
of Eye

Nasal
Septum 12

Oropharynx
13

Middle
Concha 11

4 Middle Meatus

Laryngopharynx
12

5 Cross Section
of Nasal Cavity

Inferior
Concha 10

6 Maxillary Sinus

4 Oral Cavity

5 Larynx

Inferior 9
Meatus

Phrenic Nerve
to Diaphragm
11

6 Vocal Fold
(Cord)

7 Cross
Section
of Oral Cavity

Tongue 8

7 Trachea

B. Frontal Section of Human Head

Intercostal Nerves To 10
Intercostal and
Abdominal Muscles

Spinal Cord 9

Diaphragm 8

Sphenopalatine Ganglion 7

1 Olfactory Nerve and its
Intranasal Ramifications

A. Respiratory Tract Nervous Control
of Respiration

Fifth
Cranial Nerve 6
(Trigeminal)

2 Intranasal Mucus Membrane
Reddened by Inflammatory
Hyperemia

3 Nasal Suppurative
Discharge from the Sinuses

Vagus Nerve 5

1 Mucous Membrane Reddened by
Inflammatory Vascular Dilitations

4 Swollen Palantine Tonsil

2 Purulent Sputum in Trachea and
on Bifurcation

D. Symptoms of Common Head Cold with Sinusitis

E. Cross Section of Inflamed Trachea
and Bronchi (Acute Tracheobronchitis)

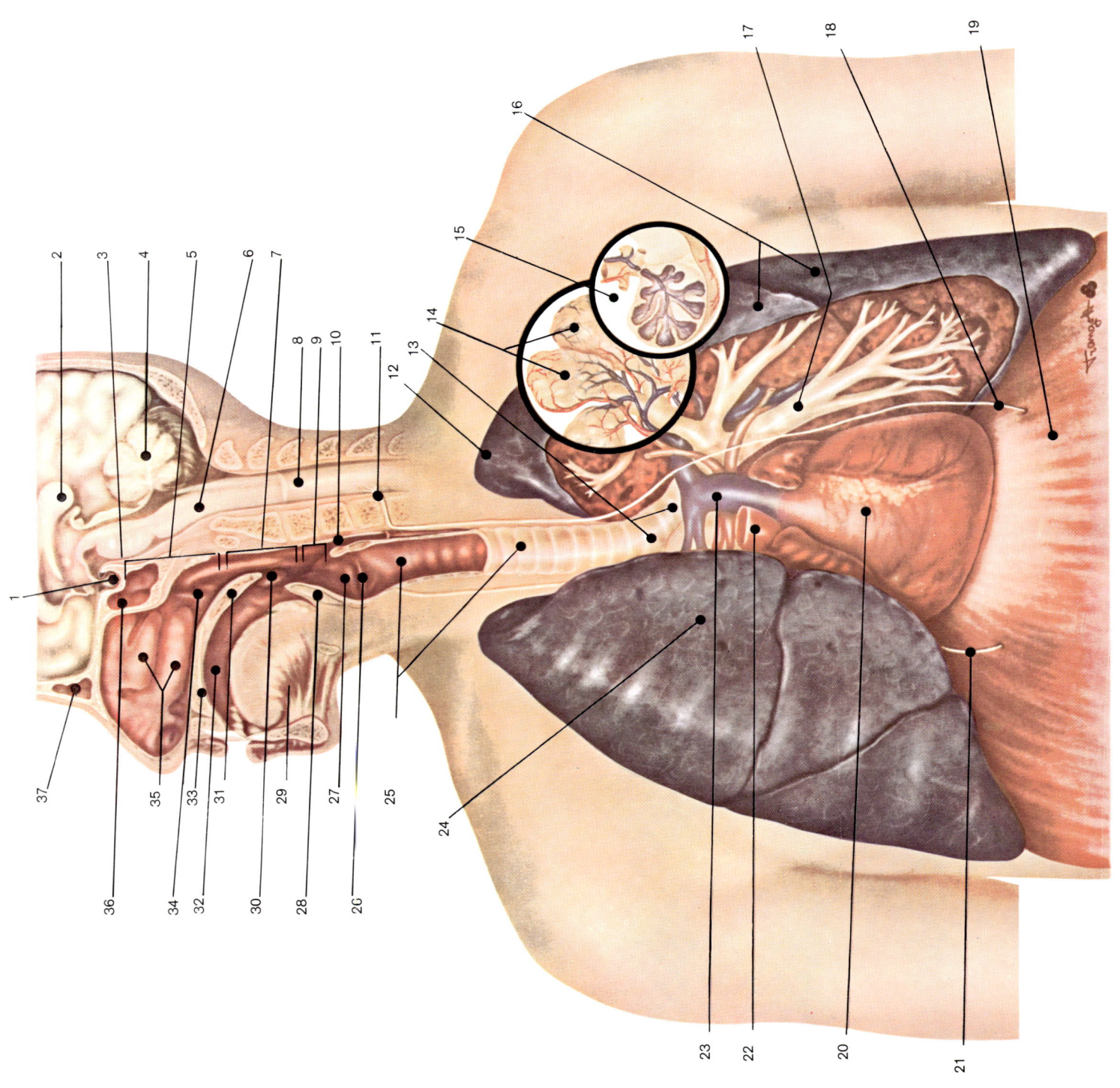

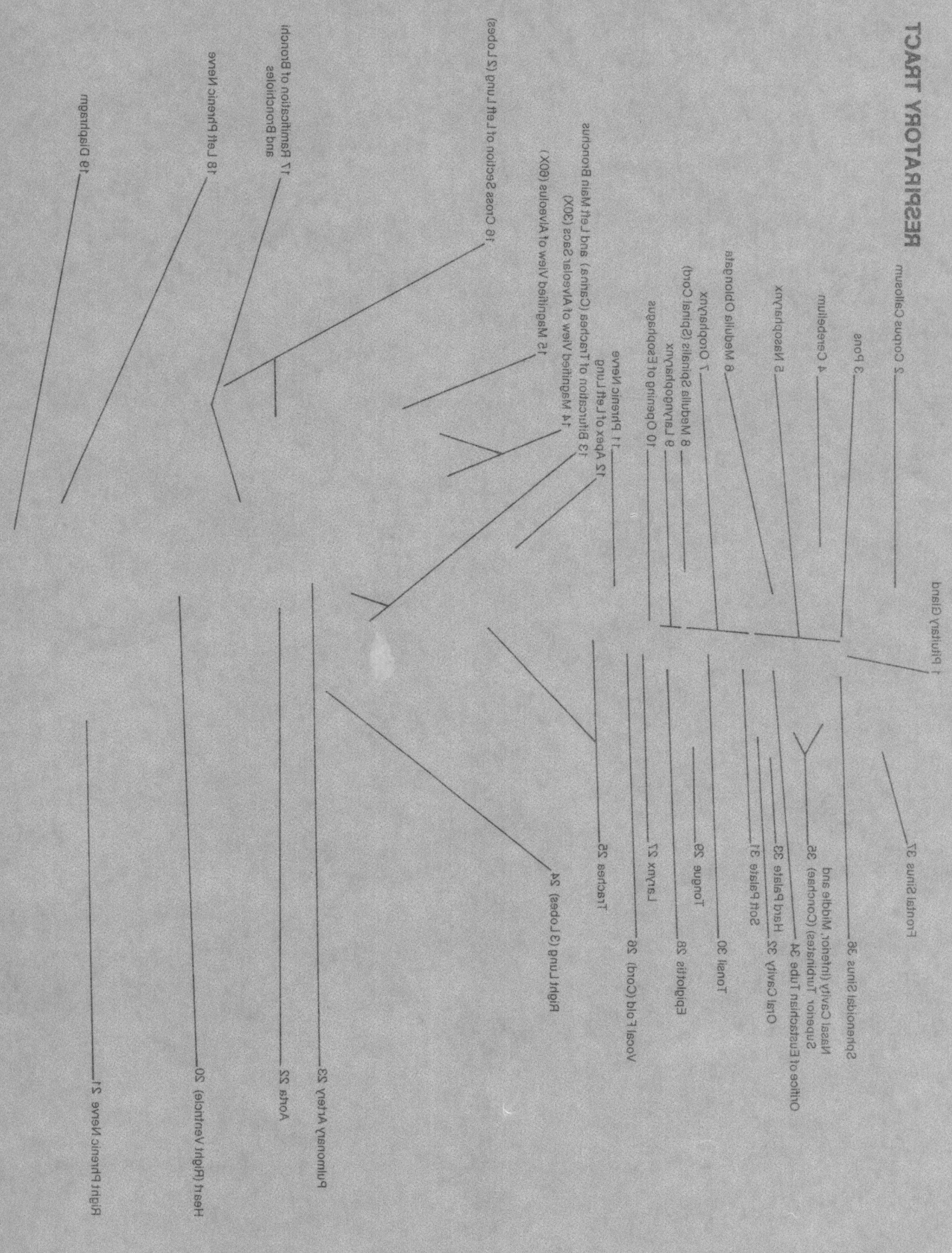

4 Cerebellum
5 Nasopharynx
28 Epiglottis
29 Tongue
30 Tonsil
31 Soft Palate
32 Oral Cavity
33 Hard Palate
34 Superior Eustachian Tube
35 Orifice of Eustachian Tube
36 Nasal Cavity (Inferior, Middle and Superior Turbinates) (Conchae)
37 Sphenoidal Sinus
Frontal Sinus 37
Left Main Bronchus
Left Main Bronchus

RESPIRATORY TRACT

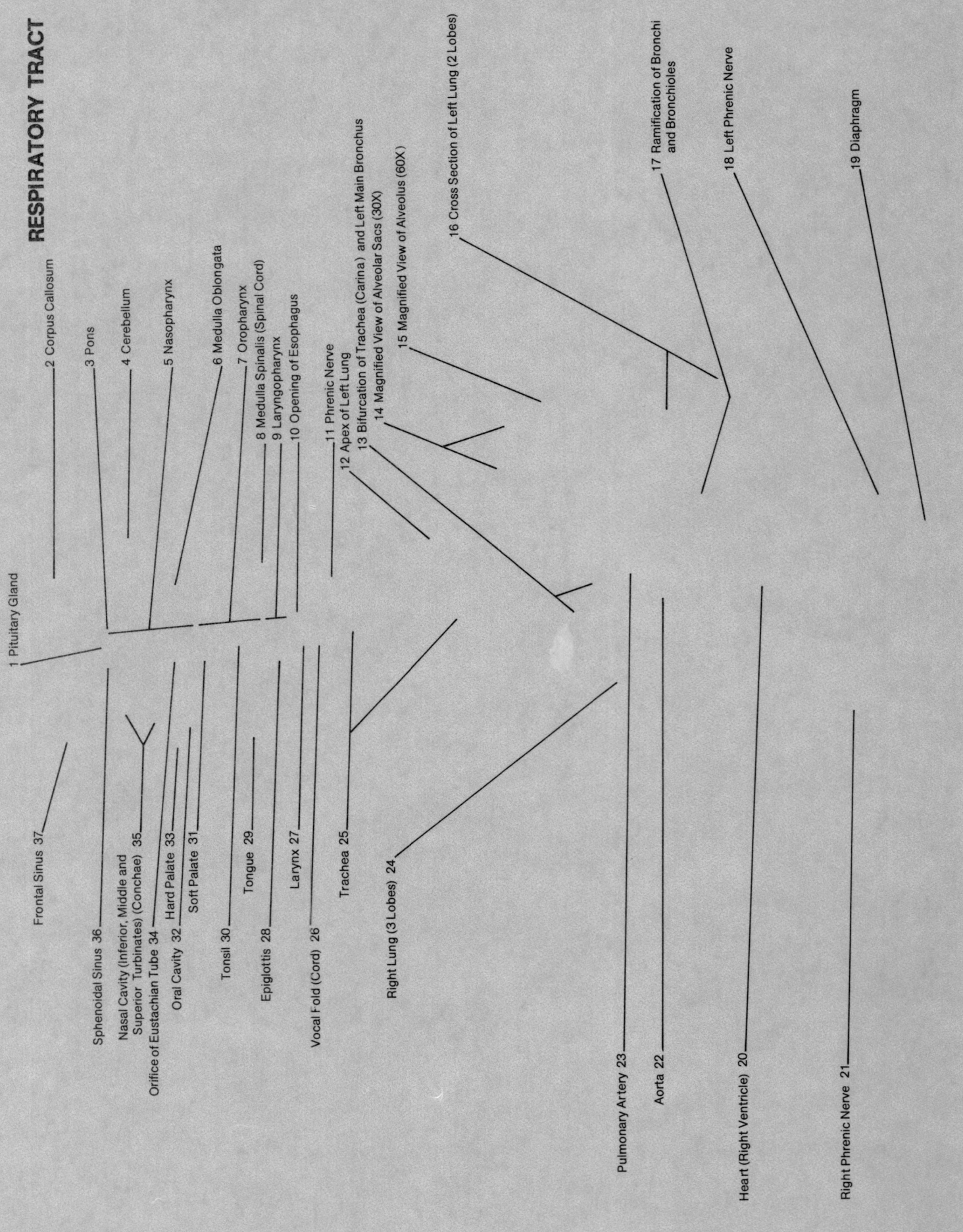

1 Pituitary Gland
2 Corpus Callosum
3 Pons
4 Cerebellum
5 Nasopharynx
6 Medulla Oblongata
7 Oropharynx
8 Medulla Spinalis (Spinal Cord)
9 Laryngopharynx
10 Opening of Esophagus
11 Phrenic Nerve
12 Apex of Left Lung
13 Bifurcation of Trachea (Carina) and Left Main Bronchus
14 Magnified View of Alveolar Sacs (30X)
15 Magnified View of Alveolus (60X)
16 Cross Section of Left Lung (2 Lobes)
17 Ramification of Bronchi and Bronchioles
18 Left Phrenic Nerve
19 Diaphragm

Frontal Sinus 37
Sphenoidal Sinus 36
Nasal Cavity (Inferior, Middle and Superior Turbinates) (Conchae) 35
Orifice of Eustachian Tube 34
Oral Cavity 32   Hard Palate 33
Soft Palate 31
Tonsil 30
Tongue 29
Epiglottis 28
Larynx 27
Vocal Fold (Cord) 26
Trachea 25
Right Lung (3 Lobes) 24

Pulmonary Artery 23
Aorta 22
Heart (Right Ventricle) 20
Right Phrenic Nerve 21

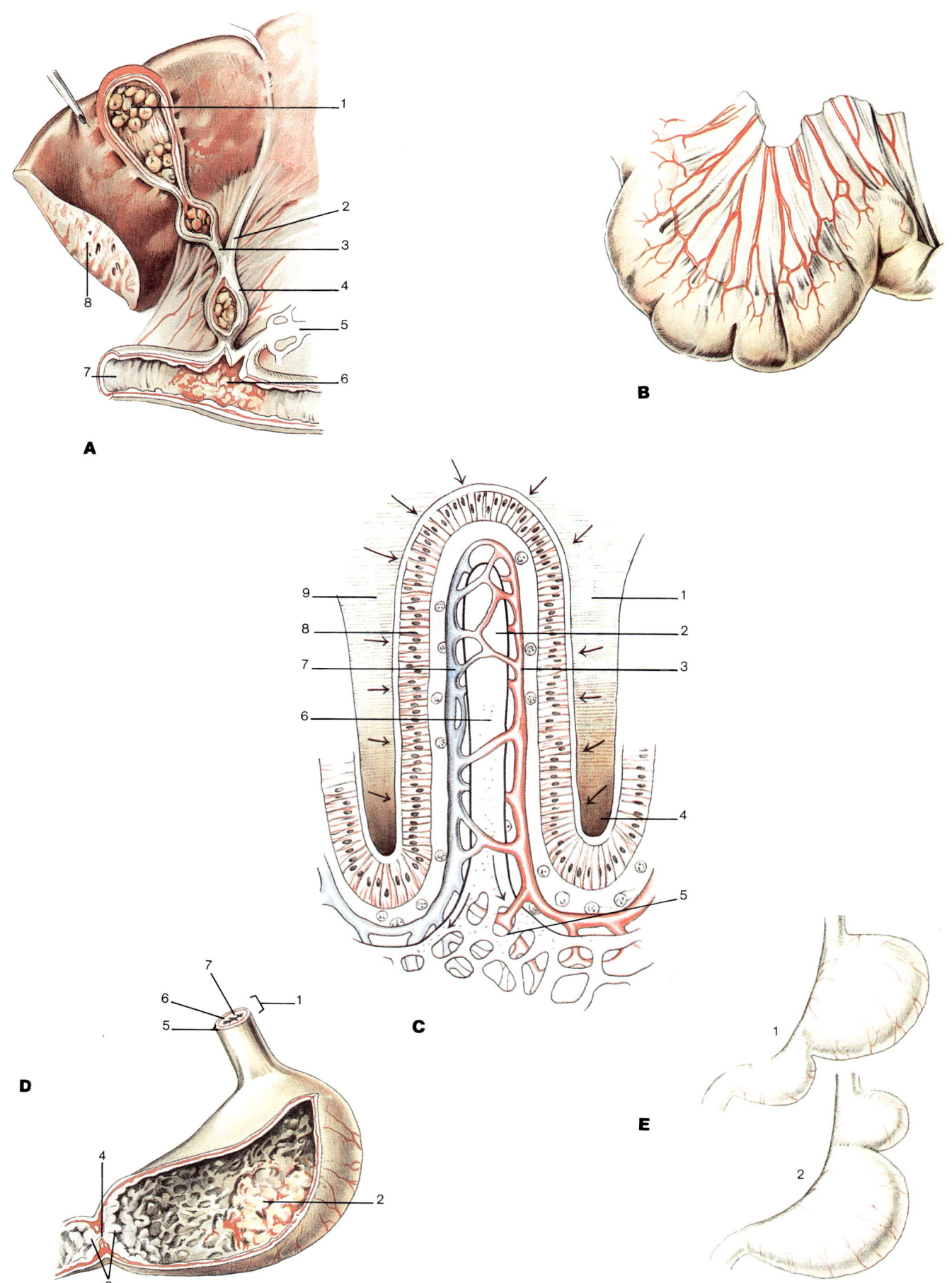

1 Gallstones _____

2 Common Hepatic Duct _____
3 Cystic Duct _____
4 Common Bile Duct _____
5 Pancreatic Duct _____
6 Duodenal Ulcer _____

B
Diseased
Liver

Duodenum 7 _____

B. Mesentery of Small Intestine

A. Liver, Pulled Upwards

1 Digested Food _____
2 Lymph Capillary (Lacteal) _____
3 Artery _____

Digested Food 9 _____
Columnar Epithelium 8 _____
Vein 7 _____

Lymph Ducts filled with Chyle 6 _____

4 Crypt of Lieberkühn _____

5 Lymph Nodule _____

C. Intestinal Villus

Mucous Coat 7
Muscle 6
Connective Tissue 5 (Serous Coat)

1 Cross Section through Esophagus

Early Constriction 1 (Stomach Filling)

D. Diseased Stomach

E. Gastric Filling and Emptying

Later Constriction 2 (Stomach Emptying)

2 Peptic Ulcer _____

Pylorus 4

3 Pylorus Deformed by Carcinoma

INTESTINAL VILLUS & DIGESTIVE DISORDERS

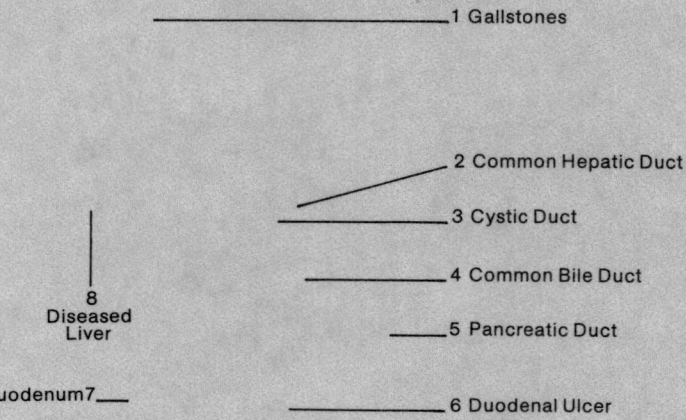

——————————— 1 Gallstones

———— 2 Common Hepatic Duct

—— 3 Cystic Duct

———— 4 Common Bile Duct

——— 5 Pancreatic Duct

8
Diseased
Liver

Duodenum 7 ———

———— 6 Duodenal Ulcer

B. Mesentery of Small Intestine

A. Liver, Pulled Upwards

Digested Food 9 ————

———— 1 Digested Food

Columnar Epithelium 8 ————

———— 2 Lymph Capillary (Lacteal)

Vein 7 ————

———— 3 Artery

Lymph Ducts filled with Chyle 6 ————

———— 4 Crypt of Lieberkühn

———— 5 Lymph Nodule

Mucous Coat 7

Muscle 6

1 Cross
Section
through
Esophagus

C. Intestinal Villus

Connective Tissue 5 ————
(Serous Coat)

Early Constriction 1
(Stomach Filling)

D. Diseased Stomach

**E. Gastric Filling
and Emptying**

Pylorus 4

Later Constriction 2
(Stomach Emptying)

———— 2 Peptic Ulcer

3 Pylorus Deformed by Carcinoma

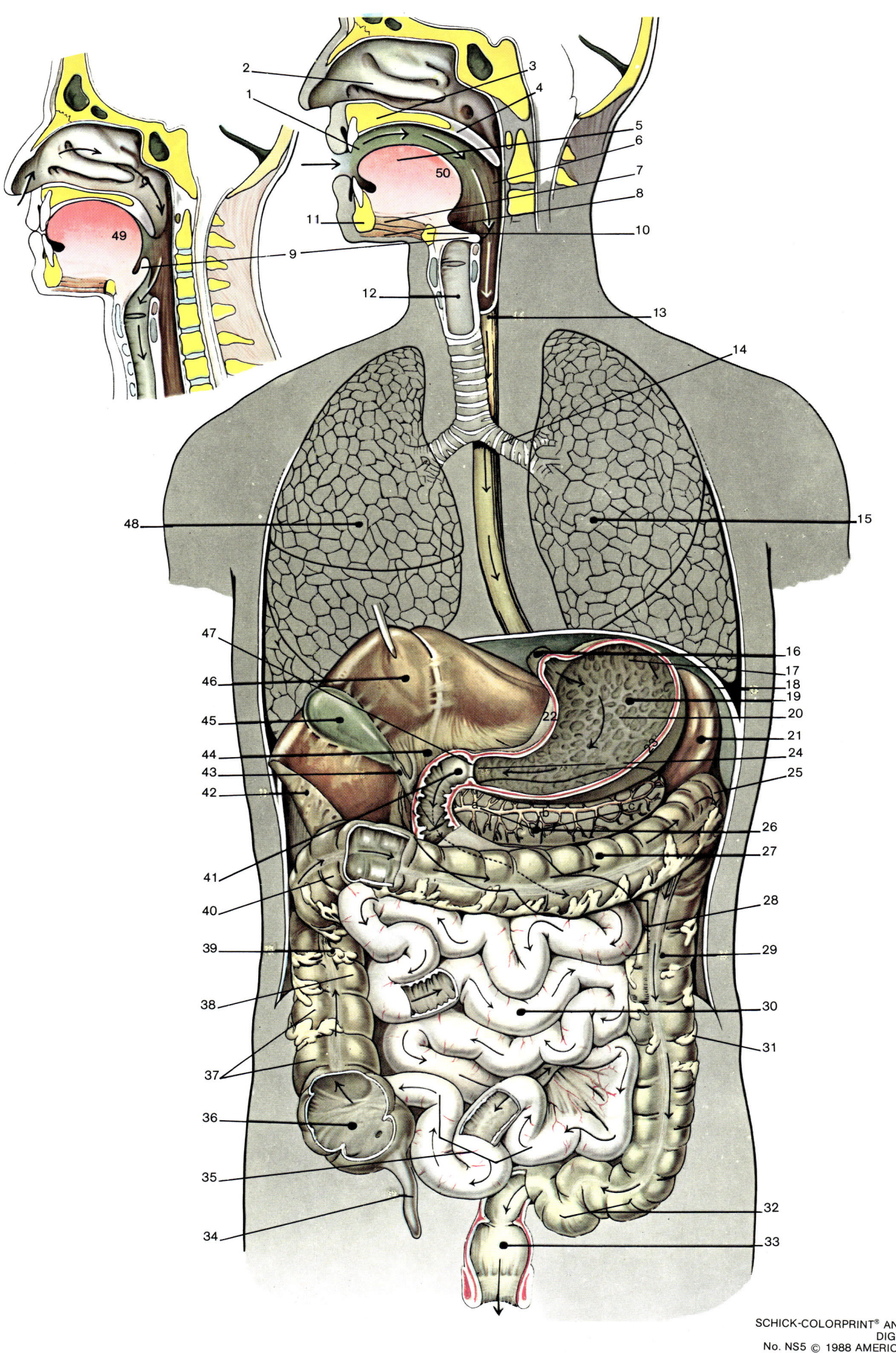

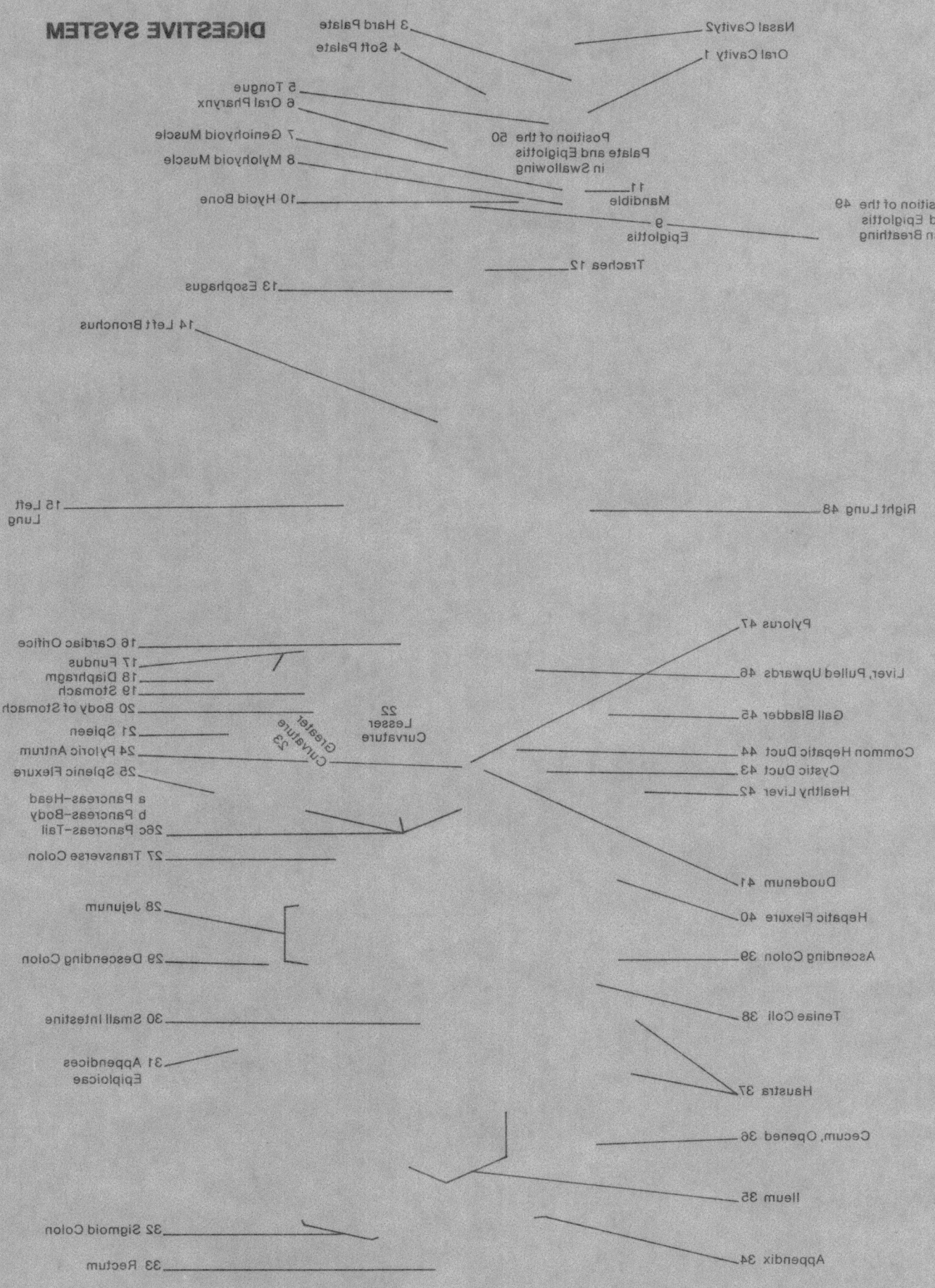

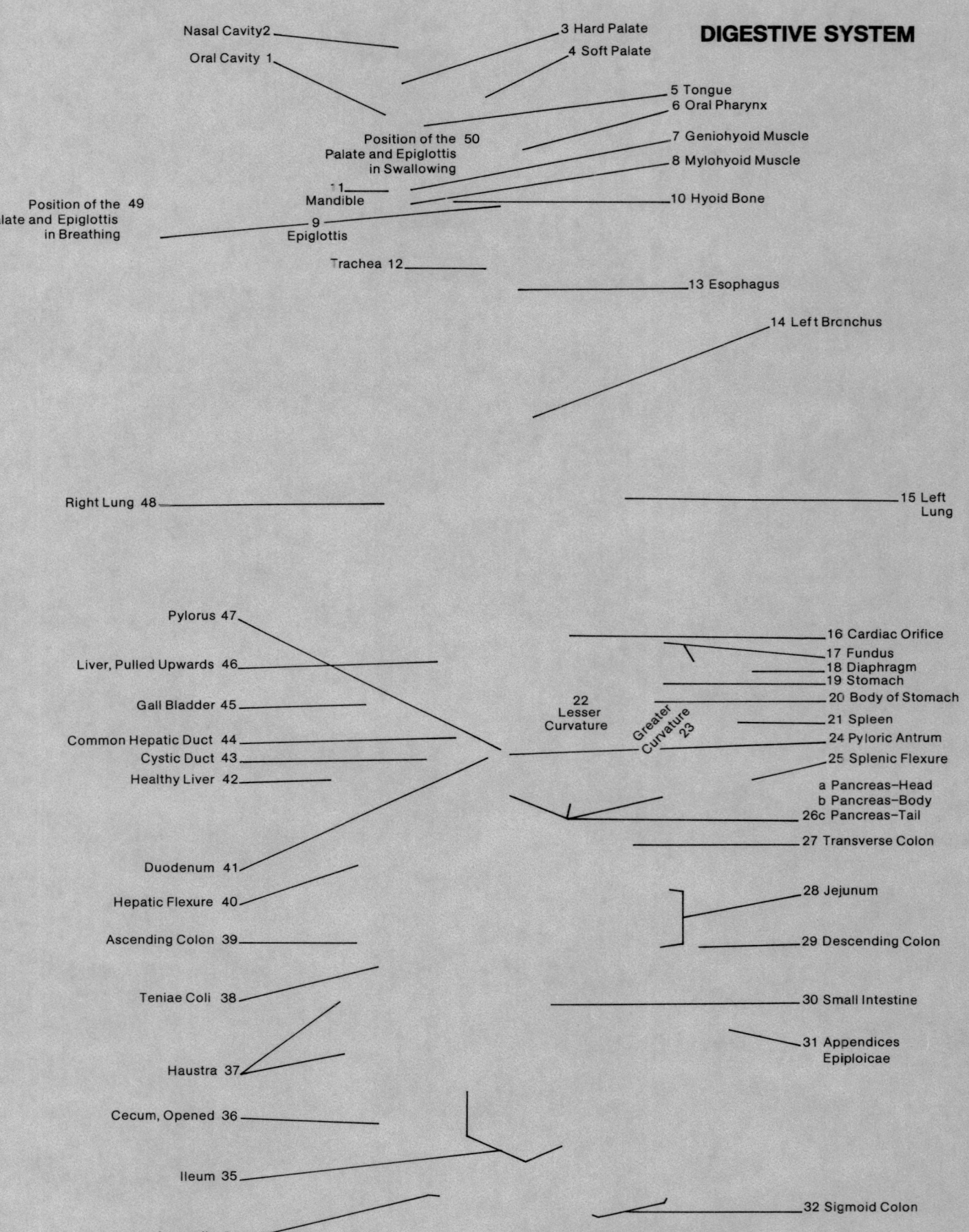

DIGESTIVE SYSTEM

Nasal Cavity2

Oral Cavity 1

3 Hard Palate

4 Soft Palate

5 Tongue

6 Oral Pharynx

Position of the 50
Palate and Epiglottis
in Swallowing

7 Geniohyoid Muscle

8 Mylohyoid Muscle

11
Mandible

10 Hyoid Bone

Position of the 49
Palate and Epiglottis
in Breathing

9
Epiglottis

Trachea 12

13 Esophagus

14 Left Bronchus

Right Lung 48

15 Left
Lung

Pylorus 47

Liver, Pulled Upwards 46

Gall Bladder 45

Common Hepatic Duct 44

Cystic Duct 43

Healthy Liver 42

16 Cardiac Orifice

17 Fundus

18 Diaphragm

19 Stomach

20 Body of Stomach

21 Spleen

24 Pyloric Antrum

25 Splenic Flexure

22
Lesser
Curvature

Greater
Curvature 23

a Pancreas–Head

b Pancreas–Body

26c Pancreas–Tail

27 Transverse Colon

Duodenum 41

Hepatic Flexure 40

Ascending Colon 39

28 Jejunum

29 Descending Colon

Teniae Coli 38

30 Small Intestine

31 Appendices
Epiploicae

Haustra 37

Cecum, Opened 36

Ileum 35

32 Sigmoid Colon

33 Rectum

Appendix 34

No. NS5 © 1988 AMERICAN MAP CORP.

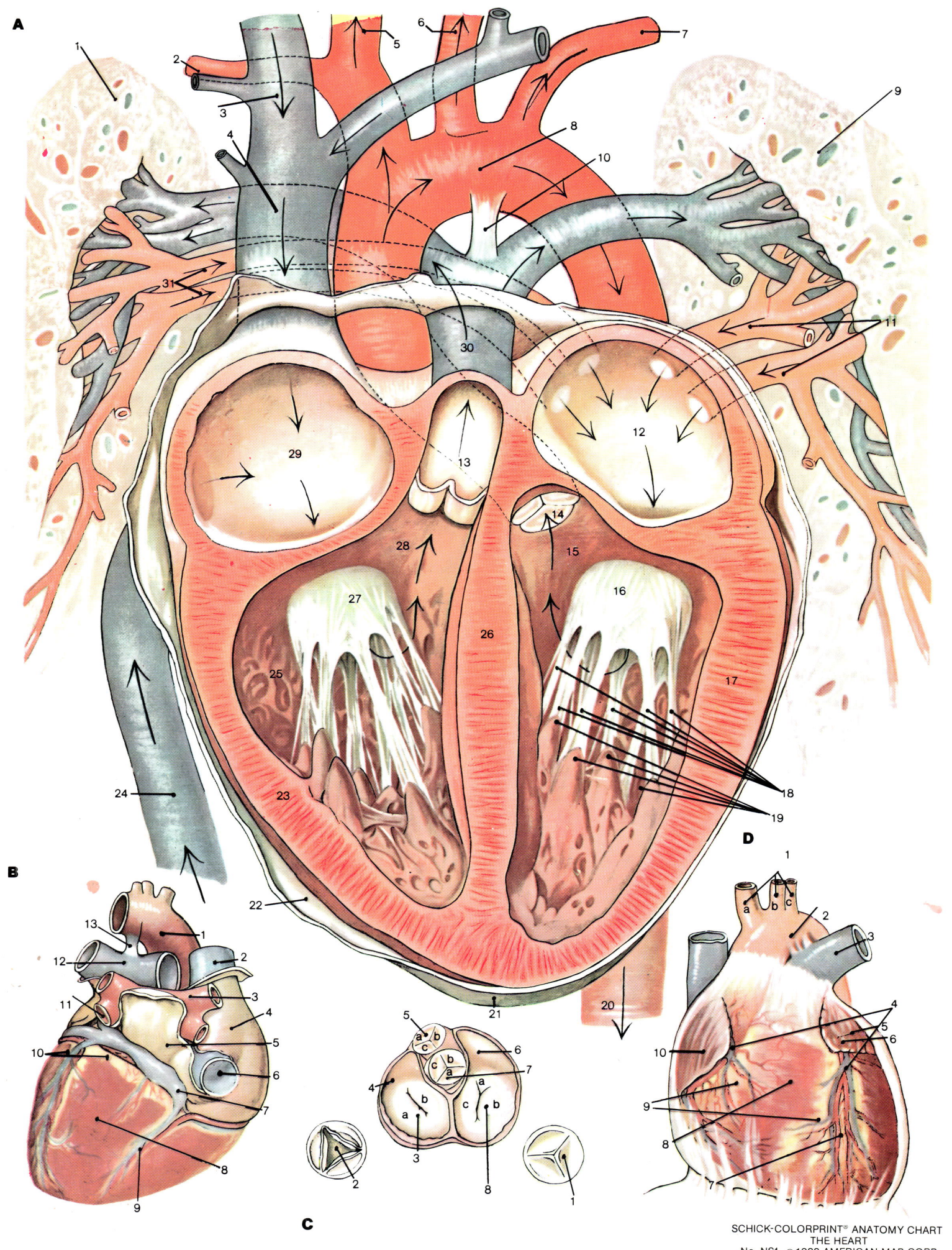

SCHICK-COLORPRINT® ANATOMY CHART
THE HEART
No. NS4   ©1988 AMERICAN MAP CORP.

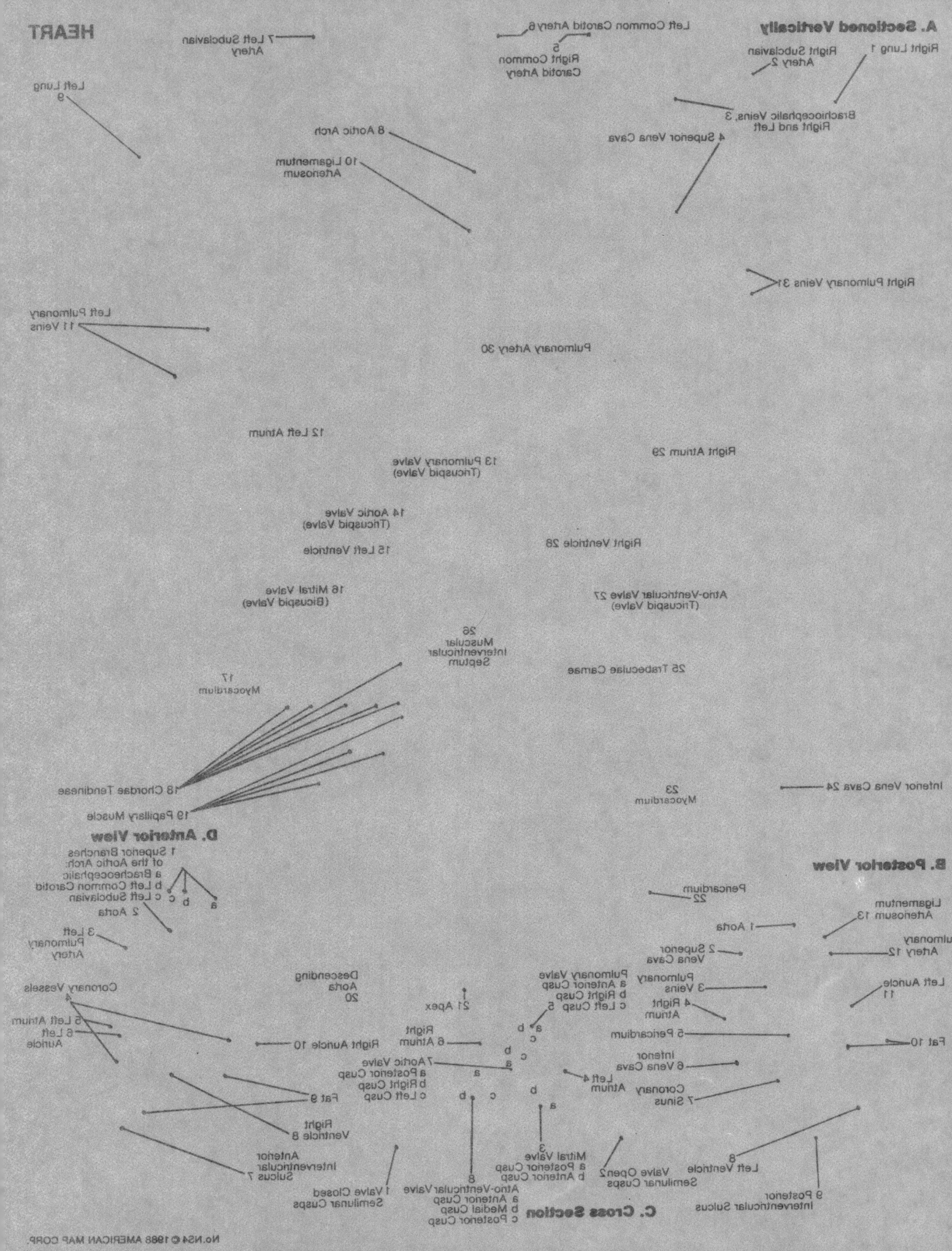

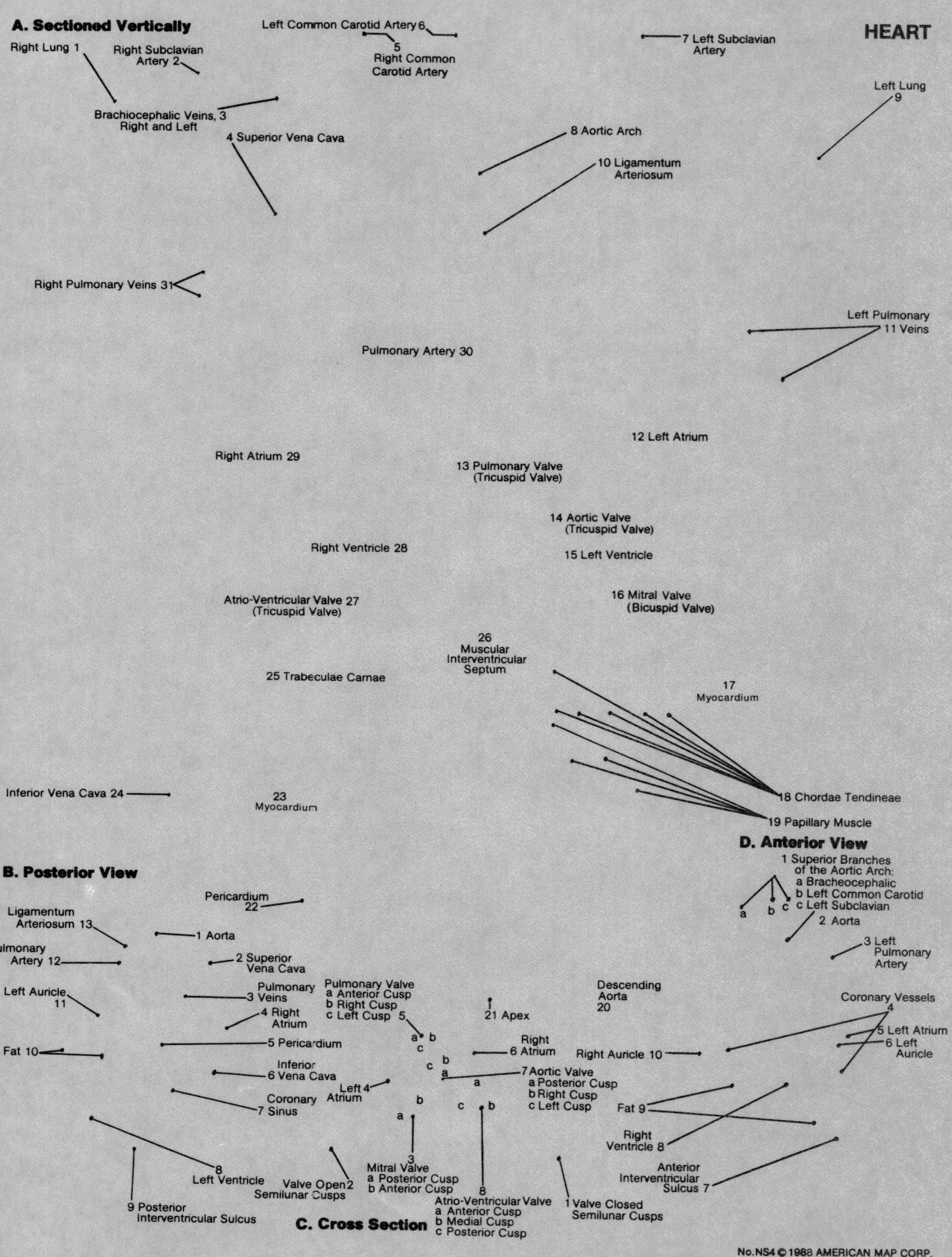

HEART

A. Sectioned Vertically

Right Lung 1

Right Subclavian
Artery 2

Left Common Carotid Artery 6
5
Right Common
Carotid Artery

7 Left Subclavian
Artery

Left Lung
9

Brachiocephalic Veins, 3
Right and Left

4 Superior Vena Cava

8 Aortic Arch

10 Ligamentum
Arteriosum

Right Pulmonary Veins 31

Left Pulmonary
11 Veins

Pulmonary Artery 30

Right Atrium 29

12 Left Atrium

13 Pulmonary Valve
(Tricuspid Valve)

14 Aortic Valve
(Tricuspid Valve)

Right Ventricle 28

15 Left Ventricle

Atrio-Ventricular Valve 27
(Tricuspid Valve)

16 Mitral Valve
(Bicuspid Valve)

26
Muscular
Interventricular
Septum

25 Trabeculae Carnae

17
Myocardium

Inferior Vena Cava 24

23
Myocardium

18 Chordae Tendineae

19 Papillary Muscle

B. Posterior View

D. Anterior View

1 Superior Branches
of the Aortic Arch:
a Bracheocephalic
b Left Common Carotid
c Left Subclavian

Pericardium
22

Ligamentum
Arteriosum 13

1 Aorta

a   b   c

2 Aorta

ulmonary
Artery 12

2 Superior
Vena Cava

3 Left
Pulmonary
Artery

Pulmonary
3 Veins

Pulmonary Valve
a Anterior Cusp
b Right Cusp
c Left Cusp 5

Descending
Aorta
20

Coronary Vessels
4

Left Auricle
11

4 Right
Atrium

21 Apex

a   b

5 Left Atrium

Fat 10

5 Pericardium

c

Right
6 Atrium

c   b

Right Auricle 10

6 Left
Auricle

Inferior
6 Vena Cava

c

7 Aortic Valve
a Posterior Cusp
b Right Cusp
c Left Cusp

Left 4
Atrium

a

a

Coronary
7 Sinus

b

c   b

Fat 9

Right
Ventricle 8

8
Left Ventricle

a

Valve Open2
Semilunar Cusps

3
Mitral Valve
a Posterior Cusp
b Anterior Cusp

a   b

8

Anterior
Interventricular
Sulcus 7

9 Posterior
Interventricular Sulcus

Atrio-Ventricular Valve
a Anterior Cusp
b Medial Cusp
c Posterior Cusp

1 Valve Closed
Semilunar Cusps

C. Cross Section

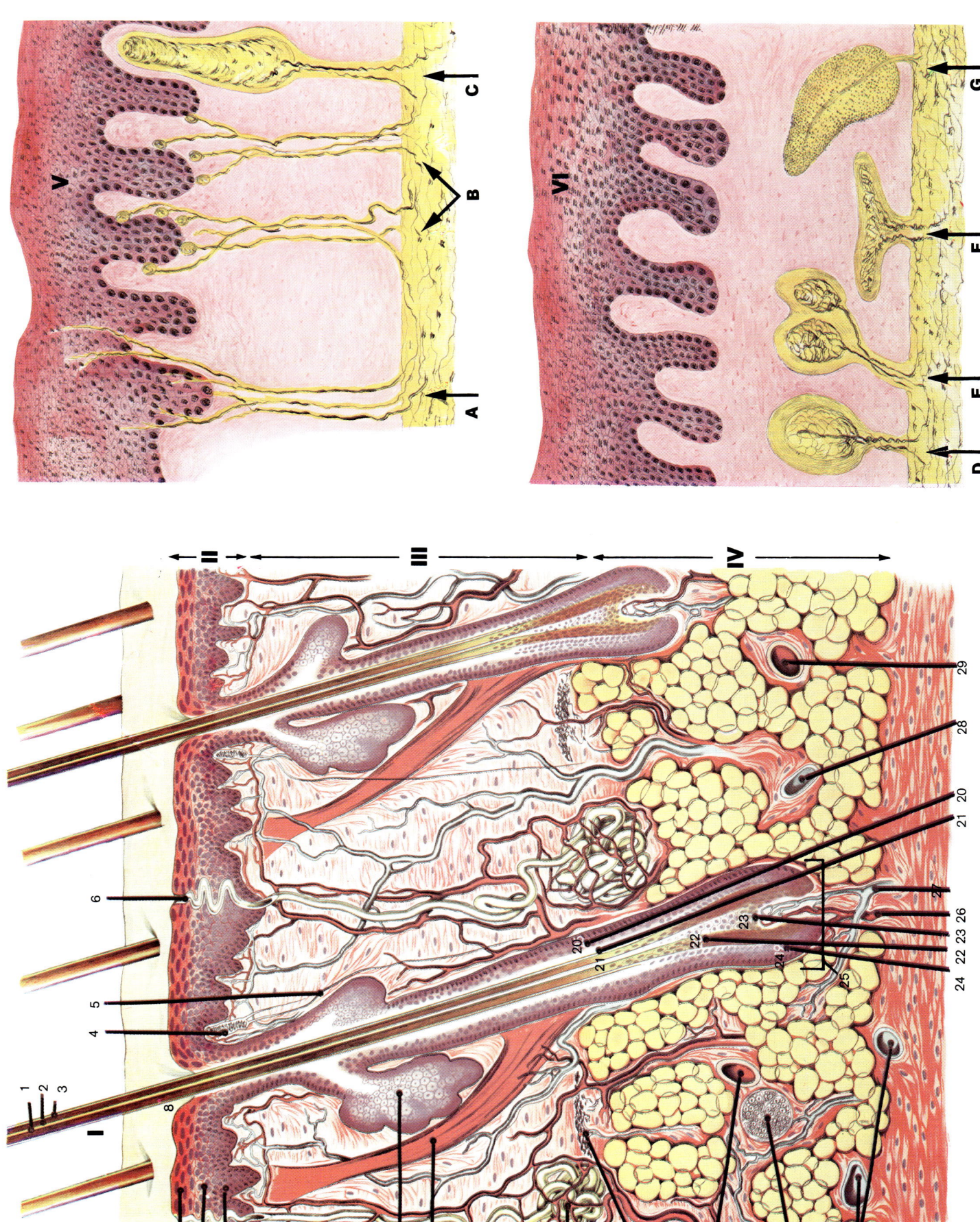

SCHICK-COLORPRINT® ANATOMY CHART
THE SKIN
No. NS3 © 1988 AMERICAN MAP CORP

SKIN

Tactile Responses V

Tactile Responses IV (Continued)

- C — Two-Point Touch discrimination
- B — Touch
- A — Pain and Temperature

- G — Traction
- F — Heat
- E — Tickling
- D — Cold

Epidermis II

Corium III or Reticular Layer

Subcutaneous Adipose Tissue IV

Hair Shaft I

Skin

1 Medulla
2 Cortex
3 Cuticle

4 Tactile Body
8 Stratum Lucidum
5 Sensory Nerve
6 Duct of Sweat Gland

9 Stratum Corneum
7 Stratum Granulosum
10 Stratum Spinosum
11 Stratum Germinativum

12 Sebaceous Gland
13 Arrector Pili Muscle
14 Spherical Body of Sweat Gland
15 Corpuscles of Ruffini
16 Artery
17 Venule
18 Cross-Section of Nerve
19 Vein

20 Dermic Coat
21 Epidermic Coat
22 Root
23 Papilla
24 Bulb
25 The Root is composed of the Bulb and the Papilla
26 Artery
27 Vein
28 Vein
29 Artery

No. N53C 1988 AMERICAN MAP CORP.

SKIN

Tactile Responses V

Two-Point Touch discrimination ← C

Touch B

Pain and Temperature ← A

Tactile Responses VI (Continued)

Cold ← D

Tickling ← E

Heat ← F

Traction ← G

Epidermis II

Corium or Reticular Layer III

Subcutaneous Adipose Tissue IV

Skin

Hair Shaft I

1 Medulla
2 Cortex
3 Cuticle

Tactile 4 Body

5 Sensory Nerve

6 Duct of Sweat Gland

Stratum Corneum 7
9 Stratum Granulosum

8 Stratum Lucidum

10 Stratum Spinosum

11 Stratum Germinativum

Subaceous Gland 12

13 Arrector Pili Muscle

Spherical Body of Sweat Gland 14

15 Corpuscles of Ruffini

16 Artery

17 Venule

18 Cross-Section of Nerve

19 Vein

20 Dermic Coat

Epidermic 21 Coat

Root 22

23 Papilla

Bulb 24

The Root 25 is composed of the Bulb and Papilla

24

22 23

26 Artery

27 Vein

21 20

28 Vein

29 Artery

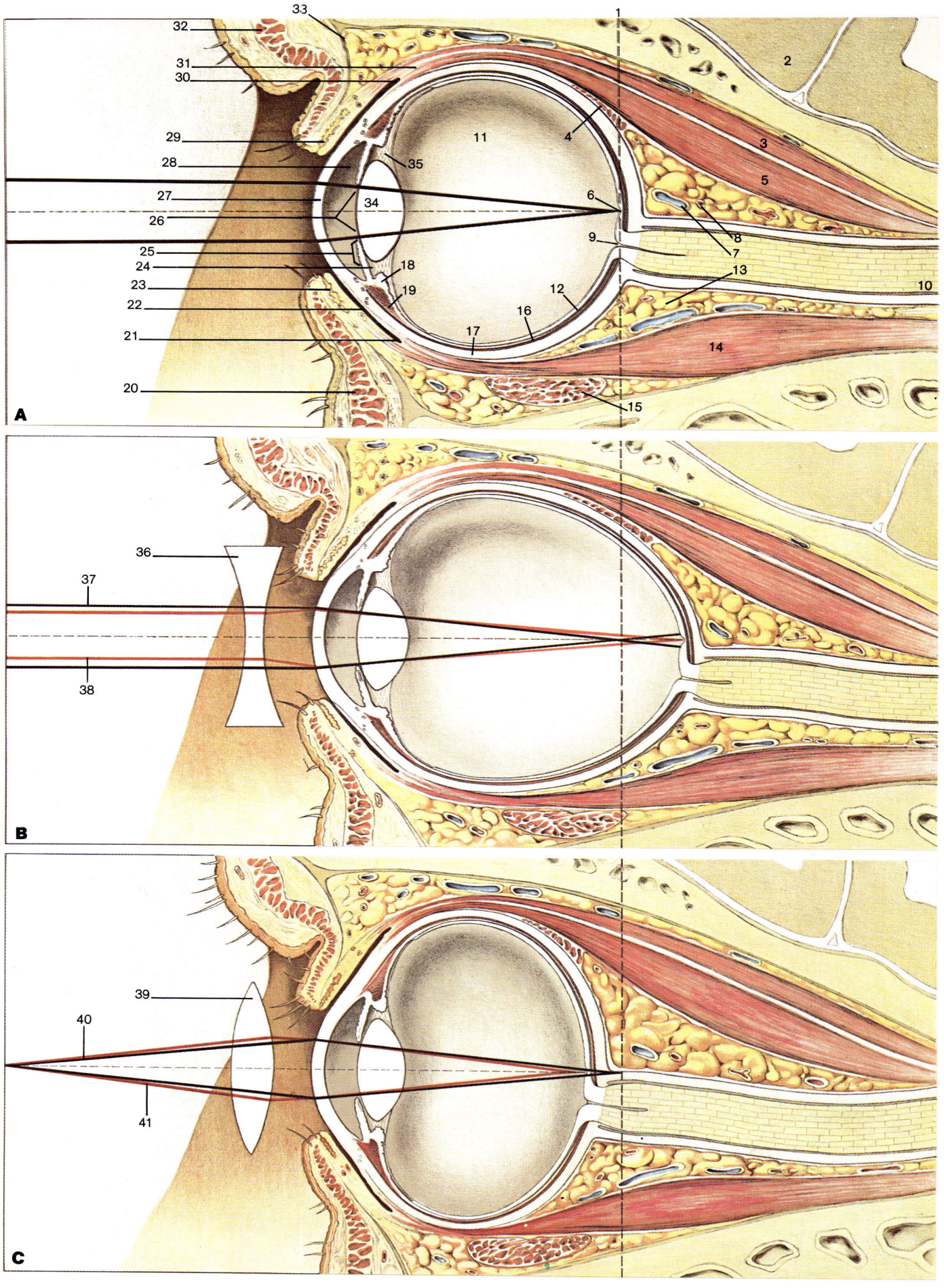

SCHICK-COLORPRINT® ANATOMY CHART
EYE-VISION
No. NS2 © 1988 AMERICAN MAP CORP.

9 Optic Papilla
8 Arteries
7 Vein
13 Periorbital Fat
10 Optic Nerve

Retina 12
Choroid 16
Sclera 17

14 Inferior Rectus Muscle
15 Inferior Oblique Muscle

A. Normal Sight

B. Nearsightedness (Myopia)

Corrective Concave Lens 36
37 Black=Without Glasses
38 Red=With Glasses

C. Farsightedness (Hyperopia)

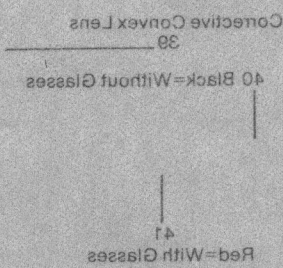

Corrective Convex Lens 39
40 Black=Without Glasses
41 Red=With Glasses

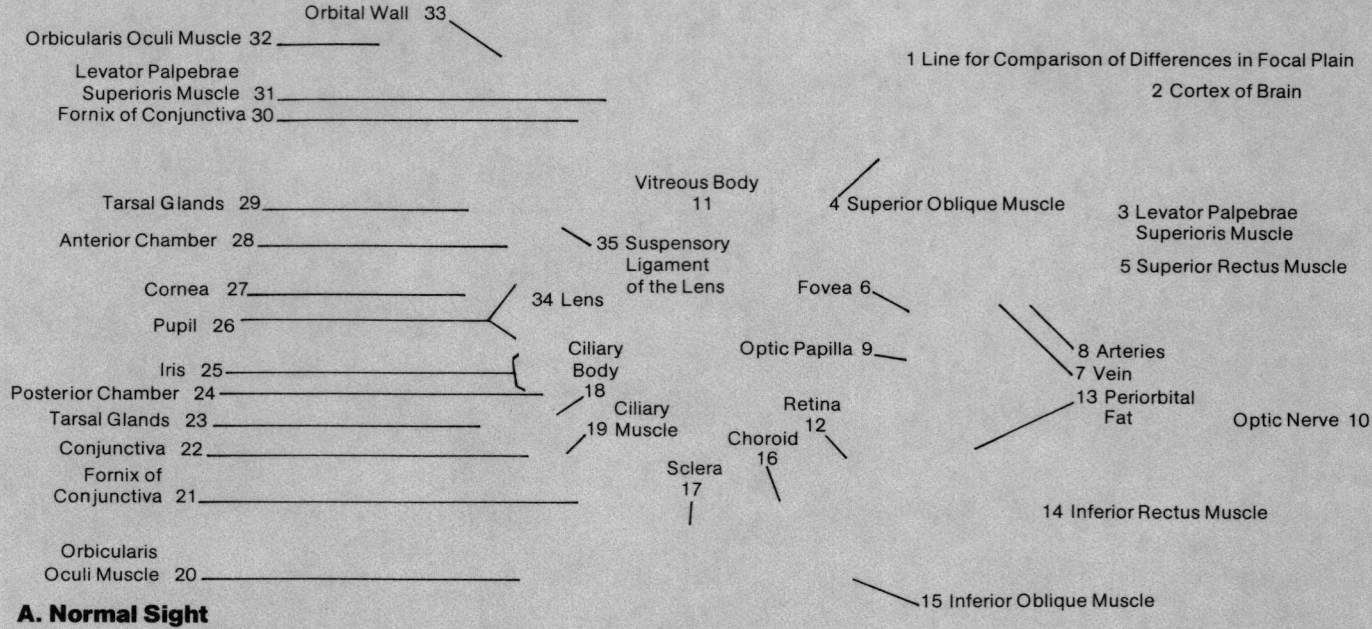

Orbital Wall 33

Orbicularis Oculi Muscle 32 _____

Levator Palpebrae
Superioris Muscle 31 _____

Fornix of Conjunctiva 30 _____

1 Line for Comparison of Differences in Focal Plain

2 Cortex of Brain

Tarsal Glands 29 _____

Anterior Chamber 28 _____

Cornea 27 _____

Pupil 26 _____

Iris 25 _____

Posterior Chamber 24 _____

Tarsal Glands 23 _____

Conjunctiva 22 _____

Fornix of
Conjunctiva 21 _____

Orbicularis
Oculi Muscle 20 _____

Vitreous Body
11

35 Suspensory
Ligament
of the Lens

34 Lens

Ciliary
Body
18

Ciliary
19 Muscle

Sclera
17

Choroid
16

Retina
12

4 Superior Oblique Muscle

3 Levator Palpebrae
Superioris Muscle

5 Superior Rectus Muscle

Fovea 6

Optic Papilla 9 _____

8 Arteries
7 Vein
13 Periorbital
Fat

Optic Nerve 10

14 Inferior Rectus Muscle

15 Inferior Oblique Muscle

A. Normal Sight

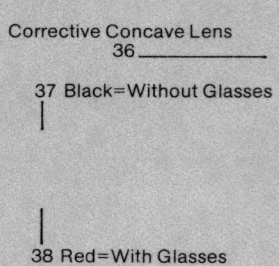

Corrective Concave Lens
36 _____

37 Black=Without Glasses

38 Red=With Glasses

B. Nearsightedness (Myopia)

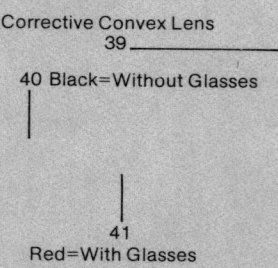

Corrective Convex Lens
39 _____

40 Black=Without Glasses

41
Red=With Glasses

C. Farsightedness (Hyperopia)

No. NS2 © 1988 AMERICAN MAP CORP.

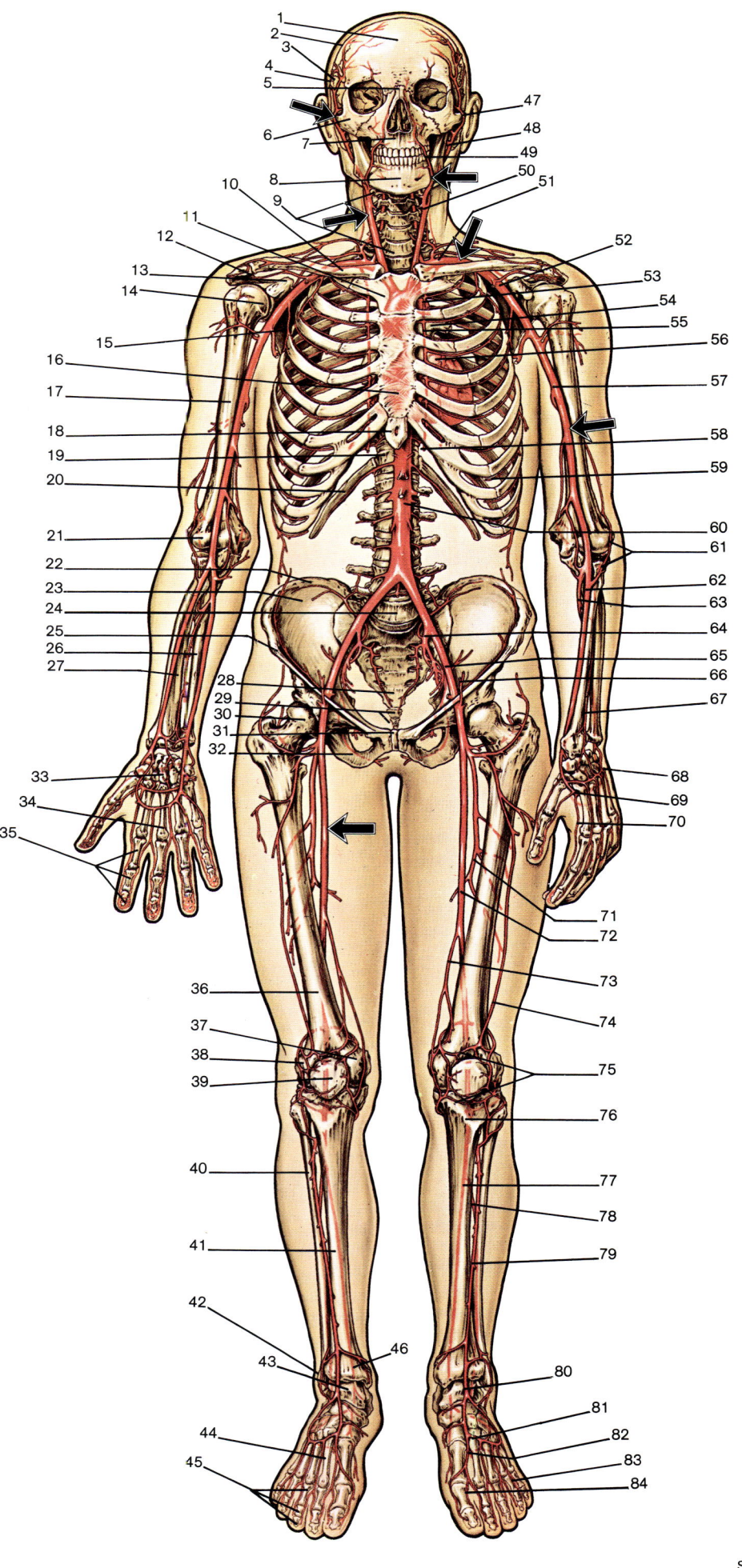

SCHICK-COLORPRINT® ANATOMY CHART
SKELETON AND ARTERIES
No. NS1 © 1988 AMERICAN MAP CORP.

Bones

1. Frontal
2. Parietal
3. Sphenoid
4. Temporal
5. Nasal
6. Arch Zygomatic
7. Maxilla
8. Mandible
9. Cervical Vertebrae
10. Clavicle
11. Manubrium of Sternum
12. Acromion (Acromial Process)
13. Coracoid Process
14. Head of Humerus
15. Scapula
16. Body of Sternum
17. Humerus
18. Xiphoid Process
19. Twelfth Thoracic Vertebra
20. Twelfth Rib
21. Condyle of Humerus
22. Iliac Crest
23. Iliac Fossa
24. Fifth Lumbar Vertebra
25. Anterior Superior Iliac Spine
26. Ulna
27. Radius
28. Sacrum
29. Coccyx
30. Head of Femur
31. Pubic Symphysis
32. Ischium
33. Carpals
34. Metacarpals
35. Phalanges
36. Femur
37. Medial Epicondyle
38. Lateral Epicondyle
39. Patella
40. Fibula
41. Tibia
42. Lateral Malleolus
43. Tarsals
44. Metatarsals
45. Phalanges
46. Medial Malleolus

Arteries

47. Superficial Temporal
48. External Carotid
49. Facial
50. Common Carotid
51. Subclavian
52. Axillary
53. Arch of Aorta
54. Pulmonary
55. Internal Thoracic
56. Heart
57. Brachial
58. Superior Epigastric
59. Intercostal
60. Abdominal Aorta
61. Anastomosis Around Elbow Joint
62. Radial
63. Ulnar
64. Internal Iliac
65. External Iliac
66. Superficial Epigastric
67. Interosseous
68. Dorsal Ulnar Carpal
69. Superficial and Deep Palmar Arches
70. Common Palmar Digital
71. Profunda Femoris
72. Femoral
73. Descending Genicular
74. Descending Branch of Lateral Femoral Circumflex
75. Circumpatellar Anastomosis
76. Popliteal
77. Posterior Tibial
78. Peroneal
79. Anterior Tibial
80. Dorsalis Pedis
81. Arcuate Artery
82. Plantar Arch
83. Dorsal Digital
84. First Dorsal Interosseus

Arrows indicate the Pressure Points of the Main Arteries.

No. N51 © 1988 AMERICAN MAP CORP.

Bones

1. Frontal
2. Parietal
3. Sphenoid
4. Temporal
5. Nasal
6. Arch Zygomatic
7. Maxilla
8. Mandible
9. Cervical Vertebrae
10. Clavicle
11. Manubrium of Sternum
12. Acromion (Acromial Process)
13. Coracoid Process
14. Head of Humerus
15. Scapula
16. Body of Sternum
17. Humerus
18. Xiphoid Process
19. Twelfth Thoracic Vertebra
20. Twelfth Rib
21. Condyle of Humerus
22. Iliac Crest
23. Iliac Fossa
24. Fifth Lumbar Vertebra
25. Anterior Superior Iliac Spine
26. Ulna
27. Radius
28. Sacrum
29. Coccyx
30. Head of Femur
31. Pubic Symphysis
32. Ischium
33. Carpals
34. Metacarpals
35. Phalanges
36. Femur
37. Medial Epicondyle
38. Lateral Epicondyle
39. Patella
40. Fibula
41. Tibia
42. Lateral Malleolus
43. Tarsals
44. Metatarsals
45. Phalanges
46. Medial Malleolus

Arteries

47. Superficial Temporal
48. External Carotid
49. Facial
50. Common Carotid
51. Subclavian
52. Axillary
53. Arch of Aorta
54. Pulmonary
55. Internal Thoracic
56. Heart
57. Brachial
58. Superior Epigastric
59. Intercostal
60. Abdominal Aorta
61. Anastomosis Around Elbow Joint
62. Radial
63. Ulnar
64. Internal Iliac
65. External Iliac
66. Superficial Epigastric
67. Interosseous
68. Dorsal Ulnar Carpal
69. Superficial and Deep Palmar Arches
70. Common Palmar Digital
71. Profunda Femoris
72. Femoral
73. Descending Genicular
74. Descending Branch of Lateral
    Femoral Circumflex
75. Circumpatellar Anastomosis
76. Popliteal
77. Posterior Tibial
78. Peroneal
79. Anterior Tibial
80. Dorsalis Pedis
81. Arcuate Artery
82. Plantar Arch
83. Dorsal Digital
84. First Dorsal Interosseus

Arrows Indicate the Pressure Points of the Main Arteries.

Contents

Colorprint® Schick
Anatomy Atlas

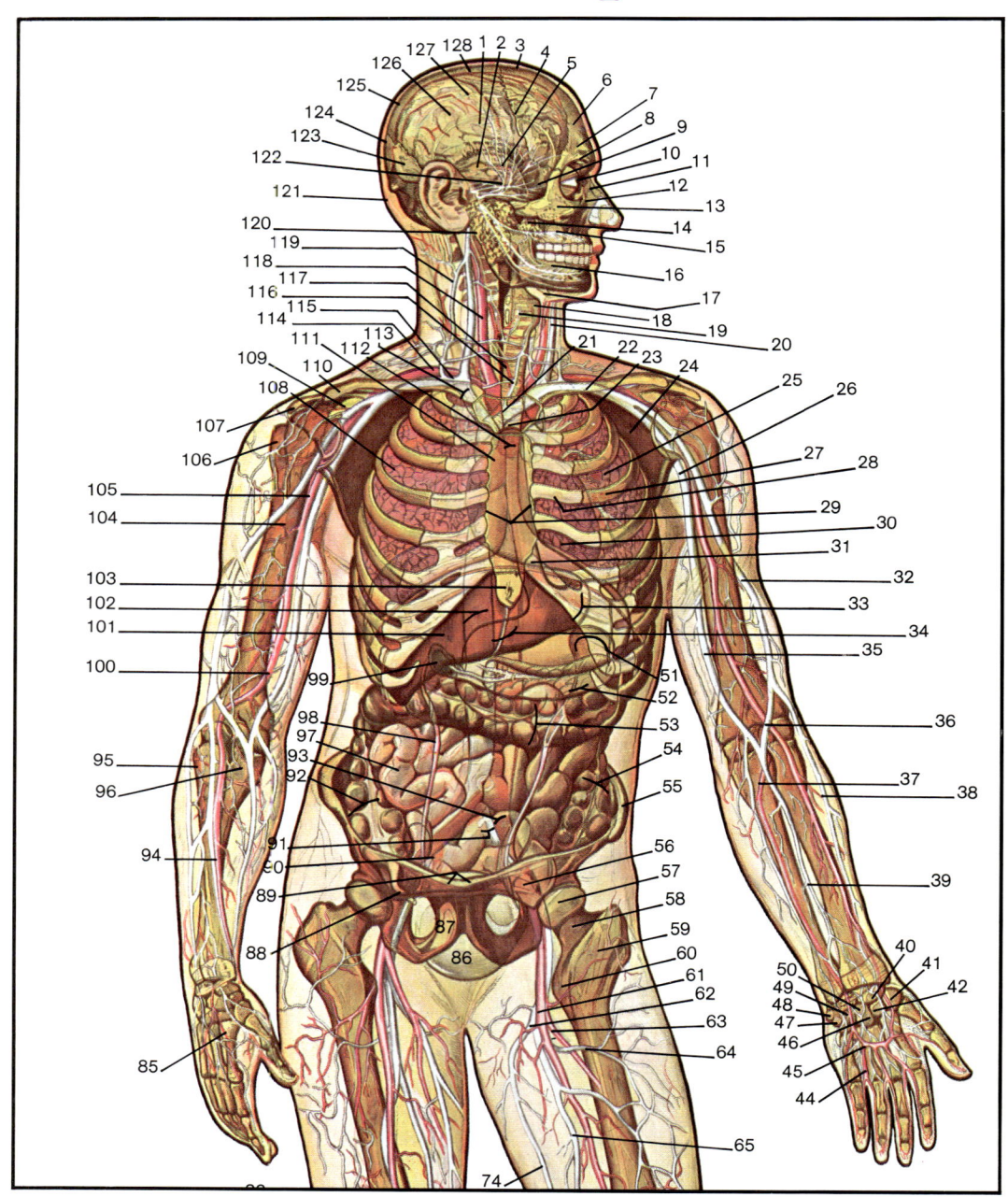

⊕American Map Corporation
46-35 54th Road, Maspeth, New York 11378